BAYOU-DIVERSITY 2

BAYOU-DIVERSITY 2
NATURE AND PEOPLE IN THE LOUISIANA BAYOU COUNTRY

Kelby Ouchley

Louisiana State University Press
Baton Rouge

Published by Louisiana State University Press
Copyright © 2018 by Louisiana State University Press
All rights reserved
Manufactured in the United States of America
First printing

Designer: Laura Roubique Gleason
Typeface: Minion Pro
Printer and binder: Sheridan Books

Library of Congress Cataloging-in-Publication Data

Names: Ouchley, Kelby, 1951– author.
Title: Bayou-diversity 2 : nature and people in the Louisiana bayou country / Kelby
    Ouchley.
Other titles: Bayou-diversity two
Description: Baton Rouge : Louisiana State University Press, 2018. | Includes index.
Identifiers: LCCN 2018027562 | ISBN 978-0-8071-6938-4 (cloth : alk. paper) | ISBN
    978-0-8071-7002-1 (pdf) | ISBN 978-0-8071-7003-8 (epub)
Subjects: LCSH: Biodiversity—Louisiana. | Bayous—Louisiana. | Species
    diversity—Louisiana.
Classification: LCC QH105.L8 O933 2018 | DDC 577.09763—dc23
LC record available at https://lccn.loc.gov/2018027562

The paper in this book meets the guidelines for permanence and durability of the
Committee on Production Guidelines for Book Longevity of the Council on Library
Resources. ∞

To my grandson, Hudson Christopher Ouchley

CONTENTS

2. Workings   64

3. Times Past   98

4. Reflections, Ruminations, and Tribulations   147

ACKNOWLEDGMENTS

Since 1995 the Monroe, Louisiana, public radio station, KEDM 90.3 FM, has aired my biweekly radio program titled *Bayou-Diversity*. As in the original *Bayou-Diversity* book, many of the hundreds of narrated essays in that series appear in this volume. I continue to appreciate this fine station's technical support and advocacy for my work.

My publishing experience with LSU Press continues to be a journey of positive experiences. In this realm, I thank MaryKatherine Callaway, Rand Dotson, Lee Sioles, and Erin Rolfs. Freelance copyeditor Stan Ivester honed my words and loaned me dozens of much-needed commas.

Emily Caldwell's talent as an artist continues to soar. I am thankful that my very good friend has agreed to share her wonderful work in this book too. I am also grateful to Gypsy Hanks, who produced the map, and to Jenny Ellerbe, who provided photography help.

My wife, Amy, continues to be a source of inspiration. It is she who deserves the most thanks.

BAYOU-DIVERSITY 2

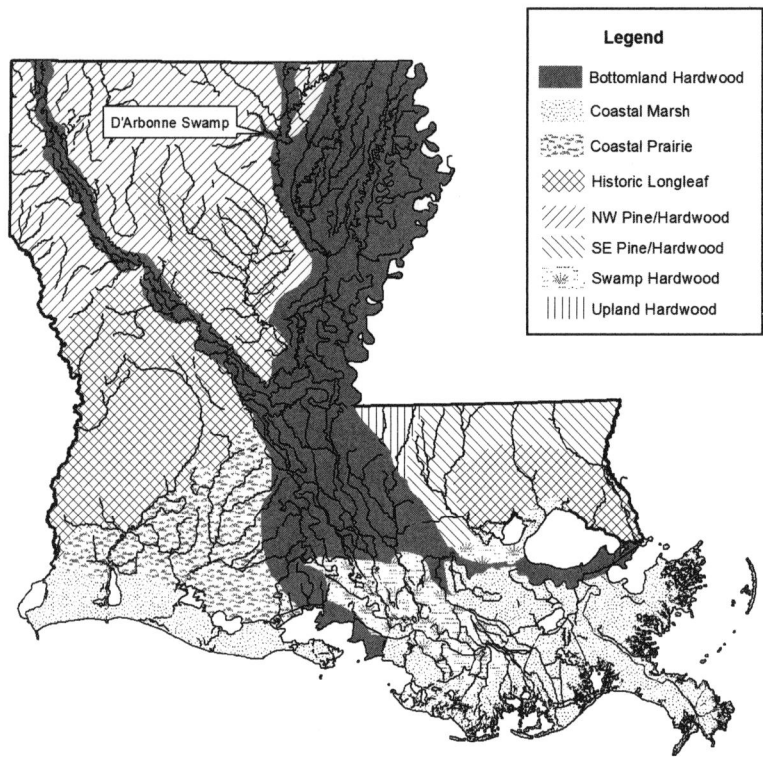

Legend

■	Bottomland Hardwood
	Coastal Marsh
	Coastal Prairie
	Historic Longleaf
	NW Pine/Hardwood
	SE Pine/Hardwood
	Swamp Hardwood
	Upland Hardwood

D'Arbonne Swamp

INTRODUCTION

When I began writing Bayou-Diversity stories for the regional public radio station more than twenty years ago, there was little talk of conservation issues in this area unless they threatened or enhanced hunting and fishing opportunities. Especially as our population becomes more urban, digital, and insulated from the natural world, there seems to be a need to provide basic information about native flora and fauna. Even more important is the necessity to educate on matters that threaten our local ecosystems and often us in the process. If there is a common thread throughout the hundreds of stories, it is that we are inextricably connected to the natural world and that our mutual well-being is inseparably linked.

Visitors to Louisiana are often perplexed by the peculiar terms we use to describe our natural features such as sloughs, swamps, and brakes. One of the most common questions asked by naïve outsiders is, "What is a bayou?" The word is ubiquitous here; after all, this is the "Bayou State." C. C. Robin, an observant traveler in the nascent days of the Louisiana Territory, tackled the definition when he wrote in 1804: "The word bayou used familiarly in this colony, and which I use frequently, does not designate, properly speaking, either a brook or a river. It is a receptacle of water which pertains to the particular conformation of this country. . . . Several of these bayous resemble rivers, and they are so multiplied in different places that travelers are often bewildered. . . . The greater part of them is filled with large trees which grow here, cypresses above all others. These trees are loaded with enormous draperies of Spanish moss, while those growing near hard water have very little or none at all. During the season of low water, these bayous which are navigable remain dry, or at the most have only a small stream of water." Today the confusion is still understandable as a broad range of businesses in the state advertise the word "bayou" in their names. We have a Bayou Bowling Alley, Bayou Builders, Bayou Forklifts, Bayou Gymnastics, Bayou Internet, Bayou Plumbing, and so forth on down through the alphabet. There are "bayou" churches, schools, and of course the fighting Bayou Bengal football team.

Early Choctaw Indians would be mystified at all this hoopla. They were responsible for the etymology of the word. French settlers took the Native Americans' perfectly good word "bayuk" and contorted it into the Franco-label "bayou." The modern definition, however, remains the same. It is a natural, relatively small waterway that flows through swamps and other lowlands in most cases. Except during flood events, currents in bayous are usually sluggish or absent. Crystalline waters are not a characteristic of bayous as they meander through heavy clay soils and capture the washed-in sediments of subtropical rains. Their shores are often dressed in live oaks and big-butted cypress trees laden with Spanish moss and parula warblers. Other riparian areas are comprised of agricultural fields of cotton, soybeans, rice, and sugarcane, or infrequently pines and upland hardwoods. Inhabitants of bayous have notable reputations and include crawfish, cottonmouths, mosquitoes, alligator gars, alligator snapping turtles, and alligators. Plenty of other species, more innocuous in their manners, are dependent on bayous and their watersheds.

Even many Louisiana citizens are not aware of the extent that bayous braid the state from north to south. More than four hundred named bayous in sixty-two of sixty-four parishes seek the shortest route to the Gulf of Mexico across a landscape void of significant relief. A few creep into the flat fringes of other Gulf Coast states and Arkansas where Bayou Bartholomew, the longest bayou in the country (about 375 miles), heads up. Bayous exist in the state's largest city of New Orleans (Bayou St. John), smallest village of Mound (Walnut Bayou), and within the corporate limits of the capital of Baton Rouge (Elbow Bayou). Glorious places if undammed, unditched, and unpolluted, they are worthy of conservation if for no other reason than Louisiana without the living gumbo of bayous would not be Louisiana, the Bayou State.

Biodiversity can be defined as all the varieties of life forms in a certain area. The area can be as large as planet earth where an estimated ten million species of plants, animals, and microbes live (95 percent of which are arthropods and microbes) or as small as a single drop of bayou water. Although the diversity of life is sometimes viewed at levels as minute as individual genes, it is more commonly considered at the scale of ecosystems. Within Louisiana several broad ecosystems are delineated by dominant vegetation types and include coastal marshes, bottomland hardwoods, prairies, pine forests, and mixed pine/upland hardwood forests. Each of these categories can be divided into more refined classifications (for example, saline, brackish, and freshwater marshes).

Species richness denotes the number of different species of plants, animals, and microbes in a given area. Different ecosystems vary in their natural capacity to support different types of life. In general, biodiversity decreases as one moves farther from the equator or higher in elevation. An important indicator of an ecosystem's health is derived by comparing current species richness against what might be expected in the area if unaltered by human disturbance. As an example, a dredged bayou draining a polluted swamp would likely have low species richness and thus poorer biological health compared to a free-flowing bayou with a pristine watershed.

The health of human societies depends on ecosystems that are species-rich by supporting processes that provide benefits to everyone. Air and water are filtered. Climate changes are moderated when forests sequester carbon dioxide. Wetlands mitigate the impacts of hurricanes and store floodwaters that could otherwise be devastating. Critical agricultural benefits are amassed in the genetic traits found in wild varieties of domestic crops. Likewise, human health benefits accrue with biodiversity as many drugs are derived, directly or indirectly, from biological origins. At least 50 percent of pharmaceuticals in the U.S. market stem from natural compounds found in plants, animals, and microorganisms. Fish, seafood, and some species of native plants and wildlife are biodiversity components that provide food and recreational opportunities. For many people the aesthetic and spiritual values of intact biodiversity are vital and immeasurable.

Bayous + Biodiversity = Bayou-Diversity. It is the variety of all living things in a place of bayous with their integral watersheds that encompass the whole of Louisiana. To a greater extent, it is the stories of every affair of natural history in the state—the stories of flora, those that produce seeds, spores, and rashes; the stories of fauna, feathered, scaled, furred, and finned. It is stories of their environment. Bayou-Diversity is also stories of human encounters and relationships with these wild things and wild places, enriching and surprising, troubling, and promising.

Readers may notice that many of the stories in this second volume of *Bayou-Diversity* have a historical bent. This is because I believe we cannot appreciate what we have gained or lost unless we understand what the natural world looked like and how it functioned before we altered it in the name of progress. Likewise, we need to have a feel of the history and culture that led to our present state of affairs. Everyone should know that there is not one acre of land or one mile of river or bayou in Louisiana, regardless of how remote, that we have not impacted. The negative effects began two hundred years ago and accumulated along with technology until today when we can look back and see the lingering shadows, some quite fresh, of industrial and agricultural pollution, drainage of our swamps, and channelization and damming of our waterways. Positive things have happened too in recent years. Tens of thousands of acres have some degree of protection in national wildlife refuges and state wildlife-management areas. In general, our streams are less polluted, but only because of better environmental regulations. Some species such as the brown pelican, black bear, and alligator that were once endangered are now thriving. It should be noted though that most of the very welcome positive conservation news is a result of our resolve to fix things that we broke in the first place. Examples include the thousands of acres of reforested wetlands in Louisiana that should never have been cleared for agriculture because they were too wet; or the dramatic comeback of our national emblem, the bald eagle, because we banned the DDT that was killing them. If only we could head off self-inflicted environmental injuries before they require expensive surgery or, worse, the patient is lost.

Bayou-Diversity 2 also contains personal anecdotes, reflections, and concerns that I think are relevant to the general theme of bayous and people. The range of stories does not easily fit into categories like those in the first book. Accordingly, the various types of articles are mixed here, and as an African safari guide often said to our group of adventurers in his lilting Swahili accent, "You never know what is just around the corner." It is my hope that the sum of stories in this book will enlighten readers regarding the remarkable biodiversity of this region, that this knowledge will be extrapolated to a scale that encompasses the planet, and that it will stimulate attitudes of proactive stewardship of the natural world.

1
FLORA AND FAUNA

Starlings and Shakespeare

That humans are increasingly impacting the natural world is beyond question. Our influence often comes by way of odd and sinuous paths. Consider that the presence of one very common but non-native and often harmful bird species in Louisiana can be attributed to William Shakespeare.

The European starling, a native of Europe and western Asia, is now found on every continent except Antarctica, a result in most cases of intentional releases by humans in the last hundred years or so. Starlings are a bit smaller than robins, with black plumage that reflects metallic green in sunlight and is covered in small white dots. Close kin to mynah birds, they have a broad repertoire of calls, including a clear, musical gurgling unlike any of our native species. They are generalists in terms of habitat preference, thriving equally well in urban and rural areas. Starlings are also not picky in their feeding habits and readily eat a wide variety of insects, spiders, worms, seeds, and fruits, items that would otherwise be available for native birds. As cavity nesters they readily commandeer bird houses and natural holes in dead trees. In areas where cavities are in short supply, this behavior may be detrimental to native birds, especially those with declining populations like the red-headed woodpecker.

The first report of a starling in Louisiana was of a single bird at

Starlings

the mouth of the Mississippi River about 1907. The next was of a small flock north of Baton Rouge in 1921. By the 1930s they were common throughout the state. These birds can be attributed to the success of a release of some eighty to one hundred starlings in New York's Central Park in 1890. As head of an obscure organization known as the American Acclimatization Society, a Mr. Eugene Schieffelin is credited with the dubious honor of bringing starlings to America. His goal was to introduce to the United States every species of bird mentioned in the works of William Shakespeare. As the bard penned in *Henry IV,* "Nay, I'll have a starling shall be taught to speak nothing but 'Mortimer'"; the destiny of starlings on the Louisiana landscape was sealed.

Flying Squirrels

Not long ago I was startled by my wife's shout to "Bring the net, hurry!" We were not fishing on the bayou bank; it was nine o'clock at night and she was in the kitchen. The perpetrator of all the excitement was a flying squirrel that had somehow managed to get inside. One thing I know: he did not fly in, as flying squirrels don't really fly—they glide.

Southern flying squirrels are common in local hardwoods. If you have large oak trees in your yard, you likely share them with flying squirrels, even if you never see them. They make dens in hollow trees, bird houses, and occasionally attics. Much smaller than gray and fox squirrels that are active during the day, flying squirrels are nocturnal. Accordingly, they have large eyes to facilitate night vision. Flying squirrels are omnivorous in that they eat a wide range of foods including nuts, fungi, bird eggs, seeds, berries, and insects. They are in turn the prey of domestic cats, raccoons, owls, and climbing snakes. Flying squirrels breed twice each year with an average litter of two or three young. Because they disperse seeds and eat arboreal insects, flying squirrels are important members of healthy, hardwood ecosystems.

Their remarkable mode of transportation is the result of a flap of loose skin that stretches from wrist to ankle on each side. When

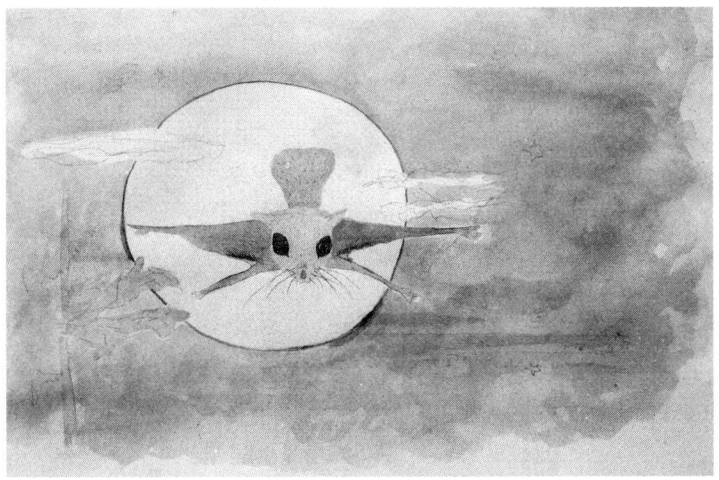

Flying Squirrel

they leap from high perches, the membranes are stretched open to serve as a parachute. A horizontally flattened tail serves as a rudder and air brake, allowing for in-flight maneuvers. From the top of a high tree they have been known to glide almost the length of a football field.

The flying squirrel in our kitchen was not happy. He bounced off of windows and cabinets. When I opened the front door, he made a beeline for the darkness outside. I stepped aside as he scurried into the calming absence of light.

Sweetgum

One of the most underappreciated native trees in Louisiana grows in every parish, is important to wildlife, and has a fascinating local history. Distinctive star-shaped leaves identify sweetgum, which grows to 150 feet tall on rich alluvial soils. During the autumn it is one of our most colorful trees as leaves on the same tree may be

Sweetgum

purple, burgundy, orange, and yellow. Sweetgum is important to several species of migrating spring warblers, each of which uses different parts of the tree to forage for insects. *Liquidambar,* the genus of the sweetgum tree, translates as "liquid amber" and refers to the waxy sap that was often chewed like chewing gum. During the Civil War, the Confederate surgeon general directed all of his medical officers to make available indigenous astringents including sweetgum for the treatment of bowel complaints among sick soldiers. Soldiers of both sides sought the plant for curative purposes. Sweetgum bark mixed with that of maple and copperas produced a purple dye, and the fruits were once used in a unique type of lighting in the South. Sweetgum balls were placed in shallow dishes filled with melted lard, and when lit the fruits produced a soft glowing light. The lustrous heartwood of large, virgin trees was known as red gum in the lumber industry. The vast Tensas Swamp in Madison and Tensas parishes was once exploited for its giant red gums by two major corporations. One area was known as the Singer Tract, where most of the wood in all Singer sewing-machine cabinets originated. The other was called the Fisher Tract and yielded lumber for Chevrolet car bodies. The last stands of the virgin trees were cut by the 1940s, but there are plenty of their offspring left to appreciate.

Carpenter Bees

Each year about the time of the spring equinox I become involved in a martial ritual that began twenty-five years ago when I first built my house in the forest. The house is made of unpainted cypress, and once vernal days lengthen to equal vernal nights it is assaulted by creatures bearing an appetency to reduce it to sawdust. The assailants are eastern carpenter bees. Their scientific name is derived from the ancient Greek word meaning "wood-cutter." My goal is to see that they cut their wood elsewhere.

Carpenter bees are often mistaken for bumblebees, but the abdomen of a bumblebee is hairy while that of a carpenter bee is shiny black. Male carpenter bees have a white face and no stinger. Even

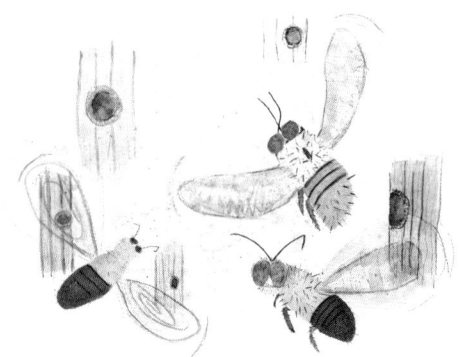

Carpenter Bees

the females, who can sting, almost never do unless provoked. My complaints are tied to their habit of building nests by tunneling into dead wood. They don't actually eat wood; instead they excavate cavities to serve as a nursery for their young. Their telltale signs are perfectly round entrance holes about three-eights of an inch in diameter. Males hover and patrol near the holes to defend their territory from other males. Eggs hatch in the summer, and the young adults often overwinter in the same tunnels. Dozens of boring bees can infest one structure.

Painted or treated wood seems to be less attractive to the bees, but preferring the look of natural wood I have spurned this route. I must say that I have learned some ways not to deter them. Plugging their holes doesn't work; they just move over an inch. The swinging motion of a badminton racket tends to inspirit them. In an instant, a .22 caliber rifle loaded with rat shot does more damage to a front-porch soffit than a carpenter bee does in a week.

After a battle encompassing a quarter-century, we seem to be at a stalemate. My new control method that involves a long-handled, fine-meshed dip net restricts them to territory already claimed. I console myself with the fact that they do have admirable traits as important pollinators for many flowers. Some, such as passion flower, can only be pollinated by the likes of carpenter bees. I'm convinced, though, that the passion of the bees is to aggravate me.

Louisiana Black Bears—"Recovered"

A few days ago, while driving through the D'Arbonne Swamp north of West Monroe, I was treated with a stunning sight. I caught a glimpse of a large animal ahead on the road shoulder, and my first impression was that it must be a hog. As I got closer it became obvious that it was a bear—shiny black and beautiful in the early morning light. He wheeled and ran down the road bank, across a shallow ditch, and into the D'Arbonne National Wildlife Refuge. Until recent years this encounter would have been nigh on impossible. The Louisiana black bear, the subspecies found in our state and historically occurring throughout Louisiana, had long been extirpated in most of its range. Once sought as a valuable commodity by French-Canadian trappers for their hides and fat, bears were later persecuted as nuisances by white settlers at every opportunity. By the time of President Teddy Roosevelt's famous 1907 bear hunt in Madison Parish, they were restricted in the state to remote areas of the Tensas and Atchafalaya swamps. Their numbers continued to decline as a result of land-clearing and illegal shooting, and by the mid-1980s when I was a manager of the Tensas River National Wildlife Refuge, our research indicated that they were uncommon even there. Concern by a broad coalition of people across the state resulted in the Louisiana black bear being formally listed under the federal Endangered Species Act in 1992. At the time the population estimate was no more than 150 bears in the state. Since then, work to restore the animals has resulted in another success story for the Endangered Species Act. The current population is estimated at 500 to 750 bears, and it was recently considered recovered and delisted—that is, removed from the Endangered Species list. Not only is this good news for Louisiana as an important part of our biological and cultural heritage has been saved, it is another success story of the often maligned Endangered Species Act. To date this forty-year-old legislation has saved 99 percent of the listed species from the brink of extinction, including our national emblem, the bald eagle, as well as the Louisiana state bird, the brown pelican, and even the icon of our swamplands, the American alligator. Would we be the same without them?

Bullbats

The award for the Louisiana bird with the most misleading name should be conferred on the common nighthawk, also known as the bullbat. No part of these monikers is accurate. In the first place, they are not common anymore as long-term surveys show their populations in the United States have declined 61 percent between 1966 and 2014. As for "night hawk," they are neither solely nocturnal nor even remotely kin to hawks. Instead, they are crepuscular, which means they are active at dawn and dusk, and they are closely related to the more familiar whip-poor-wills and chuck-will's-widows. The term "bullbat" may have roots in the courtship behavior of this species. Males perform impressive display flights of steep dives and abrupt maneuvers that produce a deep roaring sound as wind rushes over their wing feathers. To someone with a vivid imagination this might be analogous to a bellowing bull. As for the "bat" in bullbat, well that takes even more inventiveness.

Common nighthawks are about the size of a robin but appear larger because of their long pointed wings. They are mottled gray with a large, distinctive white patch on each wing and another under the chin. They are most active while feeding in the failing light on either side of sunset as they fly in ovoid loops and erratic glides. Nighthawks can sometimes be seen feeding in the glow of stadium lights and illuminated billboards that attract clouds of insects. Their in-flight call is often described as a buzzy *pzeeent*. During the day they are well camouflaged as they roost on tree branches, fence posts, or the ground. They do not construct nests and lay two eggs directly on the ground. Not uncommonly, they also choose flat roofs for nest sites. Common nighthawks are long-distance migrants nesting throughout much of the United States and wintering in South America. In Louisiana they arrive in April and depart in September. Their life on the wintering grounds is still mostly unknown to science.

The reason for their dramatic decline in recent years is unknown also. They feed exclusively on flying insects. One nighthawk had more than five hundred mosquitoes in its stomach; others had gorged on flying ants. Pesticides that eliminate these foods are suspect. One thing is certain if their population continues to collapse: we will not be discussing the aptness of their name; we'll just call them "gone."

Alligator Description

"Their toes are five in number on the anterior feet, and four on the posterior; their sharp and conical teeth are arranged in a single series in each jaw; their tongue is flat, fleshy, and closely attached almost to its very edge; and their bodies are clothed with large, thick, square scales, the upper of which are surmounted by a strong keel, those of the tail forming superiorly a dentated crest, double at its origin."

So goes the description of an alligator kept in the Tower of London menagerie in 1829. Pass these facts on to your friends and family as you eagerly await the next episode of *Swamp People*.

Alligators are reptiles in the taxonomic class called Reptilia. Members of this group have common characteristics. All are ecto-thermic or cold-blooded and have backbones. Most have four limbs (except snakes, which have four-limbed ancestors), reproduce by laying eggs with shells (except, again, for some snakes), and have bodies covered in scales or scutes. Within Class Reptilia, alligators are placed in the subdivision known as Order Crocodilia and are referred to as crocodilians. Members of this group have similar anatomical traits and include two species of alligators, thirteen kinds of crocodiles, six species of caimans, and the gharial. Alligators differ from crocodiles by having a broader snout and an upper jaw that overlaps teeth in the lower jaw. The gharial has a long, slender snout. Alligators most closely resemble caimans that live in Central and South America. The American alligators that Troy Landry dramatically pursues grow larger than their closest relative, the Chinese alligator, which rarely exceeds seven feet in length. Still, there might be potential in an oriental version of *Swamp People* for any of you entrepreneurs out there with Beijing connections.

Cottonwood I

A saddled horse standing beside a giant eastern cottonwood is the subject of a nitrate-based cellulose negative given to me by the man who took the shot in 1938 while prowling about for ivory-billed woodpeckers in Louisiana's vast Tensas Swamp. The tree appears to be nearly as wide as James Tanner's sorrel gelding is long. Even in what then was the closest thing remaining to a large, old-growth bottomland hardwood forest in America, the tree in its size was an anomaly. Why else would Tanner, who had encounters there with panthers, wolves as shiny black as the back of a mud snake, and the mythical woodpecker that he actually held in his hands, waste precious film on a big tree?

Cottonwoods were minor components of eastern forests. In the lower Mississippi Valley their prime haunts are still the front land ridges of the batture, terrain now between the levees and the rivers. There, on relocated silt that once nourished big bluestem in mid-continent prairies, cottonwoods sometimes grew 190 feet above the sandbars.

Tanner's tree, typical of the species, represented disturbance. As a Goliath it also reinforced misconceptions concerning "virgin" forests. Such a giant buttressed romanticists' oneiric ideas of primeval nature for a school of believers. In fact, cottonwoods are one of the fastest growing but short-lived eastern hardwoods. Tanner's tree germinated when one of millions of seeds from a single female tree on the high bank of the Tensas River hit pay dirt after drifting under its cotton-like parachute into a recently abandoned plantation field.

In a historical context, General Grant, in his occupation of the area while trying to subdue Vicksburg, deserves full credit for providing the indirect disturbance necessary to kick-start Tanner's cottonwood. Obscuring almost all traces of an economy that resulted in war, a functional forest that included Tanner's tree soon returned to the antebellum plantation sites. Today, only brick-lined cisterns, haunting cemeteries with toppled tombstones, and a gin chimney lost in the remote swamp survive as reminders of the latent seedbeds awaiting giant cottonwoods and other opportunists.

Cottonwood II

Louisiana residents have never associated cottonwood trees with matters of life and death. Large fast-growing trees that often exceed a hundred feet in height, they are most common in the new soils of batture wetlands along rivers and bayous. As an invader with wind-blown seeds, cottonwood is an early succession stage plant that is often the first to colonize sandbars and abandoned riparian fields. Rabbits, beaver, and deer relish the twigs and bark, but humans in Louisiana have found little use for the tree other than as inferior wood for boxes and crates. On a small scale, wild trees and those planted in plantations are used for pulpwood.

In other parts of the country, cottonwoods have been venerated for millennia. They are the dominant structural component of riparian ecosystems on the Great Plains and other areas of the arid West. Providing shade and food for a host of species, cottonwoods along streams are often the only tree-size plant on millions of acres. Native Americans, especially those in the desert Southwest, knew the presence of cottonwoods as indicators of life-sustaining seeps and springs. The Anasazi built their enigmatic cliff and canyon dwellings adjacent cottonwood-rimmed springs and guarded them with adobe watchtowers in a survival-of-the-fittest environment.

Later, when Euro-Americans marched across the continent with diurnal dreams of Manifest Destiny, the western cottonwoods themselves were besieged. As the only available fuel wood, many were fed to the voracious boilers of shallow-draft steamboats. Two centuries of dam-building disrupted natural flood cycles critical for the establishment of cottonwood seedlings, millions of cattle consumed the remaining sprouts, and competitive invasive species such as tamarisk moved in. Thus stressed by a perfect storm of unnatural conditions, few young trees are growing up to replace the old cottonwoods.

Recently, my wife and I sought reprieve from the mid-August furnace of a southeastern Utah desert. We found it in the shade of a large Fremont cottonwood in a Bureau of Land Management campground on the San Juan River, a stream less impressive in size than Bayou D'Arbonne. We sat there through the heat of the afternoon looking out over scattered juvenile cottonwoods recently planted and constantly irrigated with great care in an effort to restore this species to its rightful place in the natural scheme of life. That night our shower consisted of sitting on a camp stool by a lone hosepipe and pouring buckets of tepid water over each other, the same water that sustained the cottonwoods. Three hundred feet behind us on a moonlit cliff, Kokopelli danced and played a flute not so much for us as for the trees that signaled water and life instead of death, at least for a while.

Belted Kingfisher

In teaching kids how to fish, one of the first obstacles that must be overcome is what has to be an innate urge to throw rocks and sticks into the water for the sheer joy of it. Every fisherman knows that such commotions only scare the fish away. Within the world of birds, though, some of the best fishers actually incorporate this behavior into their pedagogy.

Anyone who has spent much time on Louisiana's rivers and bayous has heard the rattling call of kingfishers and seen their undulat-

Kingfisher

ing flight reflected from dark waters as they disappear around the nearest bend. Up close they are powdery blue jewels with ragged crests and stout tern-like bills. Males have a white breast spanned by a blue band. Females have a blue and a chestnut band. The pigeon-sized belted kingfisher is the species common in Louisiana.

As the name suggests, their diet is primarily fish, thus tying them to various aquatic ecosystems such as those that encompass rivers, bayous, lakes, bays, or coastlines. They cannot survive in areas where the water is frozen for long periods, making Louisiana ideal year-round habitat. The state's resident population of kingfishers is augmented in winter by northern birds seeking ice-free feeding areas.

Another obligate habitat requirement during the breeding season is vertical, earthen banks for nesting. Kingfishers excavate upward-sloping burrows several feet into soft banks where the female lays five to eight eggs. The banks are often over or adjacent to water. Both sexes incubate the eggs with the female usually taking the night shift. The eggs hatch after about three weeks, triggering frenetic fishing expeditions by the parents to feed the ravenous young birds.

In addition to small fish, prey occasionally includes insects, crawfish, frogs, and tadpoles. Kingfishers hunt by plunging into the water from a perch or by hovering over prey before diving in, bill downward to capture quarry at or just beneath the surface. After a dive, whether successful or not, they often fly away emitting their loud, rattling cry.

The conservation status of belted kingfishers is that of a species in decline. The reasons are unclear, but one suspected cause is tied to the loss of critical, vertical banks that are subject to erosion and near water for nesting. Such habitat is often impacted by levees, and navigation and bank-stabilization projects.

As for the business of teaching youngsters to fish, kingfishers break the rules. Young kingfishers apparently don't have a strong, natural proclivity to dive head-first into murky waters in pursuit of dinner. Adults tutor them by dropping dead fish into the water for their retrieval. And as counter-productive as it may seem to us, they have also been observed dropping sticks in the water for the same purpose.

Spotted Bass

Always in late February when the first white crawfish reached two inches in length a ritual began in the D'Arbonne Swamp that included my father, his cousin, and me, an adolescent youth in those years a half-century ago. The object of the tradition was to procure "smallmouth bass" for the deep, black skillet. Using a long-handled, homemade drag consisting of a joint of one-inch galvanized pipe attached to a basket of hardware cloth, we scraped the ditches along White's Ferry Road for the tender crawfish. When a sufficient number filled the steel, five-gallon bucket stuffed with Spanish moss, we were set to go the following morning. Barring a late cold front, we launched the low-sided Arkansas Traveler at Holland's Bluff landing at first light as wood ducks beat their way from buttonbush roosts to the oak flats. The bayou was different then before the Corps of Engineers' latest edition of navigation improvements on the Ouachita River drowned the shallow, rippling gravel bars under quiescent, sediment-laden pools. Dad knew the location of these unique habitats even if they were hidden under winter backwater, and the seven-and-one-half-horsepower outboard pushed us upstream toward Old Mill, site of a short-lived sawmill, or perhaps the wreck where the steamboat *Tributary* burned and sank in 1890. Here we anchored, fitted spin-cast rigs with lead weights and 2/0 hooks baited with the crawfish, and flung the offerings into the cold, dark bayou. Almost always we caught the spunky red-eyed bass.

Many years and a challenging ichthyology class later I learned in no uncertain terms from a favorite professor that smallmouth bass are not native to Louisiana because they are adapted to live in cooler waters. The bass that we caught then on Bayou D'Arbonne are properly called spotted bass, members of the black bass group along with largemouth and true smallmouth bass. Other common names of the spotted bass are "Kentucky bass" and "redeye bass." Resembling slender largemouth bass, spotted bass have black splotches along their sides, and indeed their mouth is smaller than that of their abundant largemouth cousins. We rarely caught a spotted bass of more than two pounds, and the Louisiana state record for the species is less than five pounds.

Spotted bass are still found throughout the state but are likely much less abundant than before most of our rivers and bayous were altered by dams, navigation projects, and pollution. They tend to be found in areas with more current than largemouth bass and usually choose gravel or rocky areas as spawning sites, habitats that are uncommon in many Louisiana streams. Reeled up from the depths of a mysterious, unseen realm beneath the surface to leap and splash in the still-angled, winter sunlight, they seemed to me inspirited treasure with fiery red eyes.

River Otters Then and Now

What do a small meandering bayou in north Louisiana and a main branch of the Mississippi River that empties into the Gulf of Mexico have in common? It seems that both were named by French explorers for a semiaquatic mammal that was abundant in each of the areas. Bayou de l'Outre in Union Parish and Pass a Loutre in Plaquemines Parish were named for river otters.

River otters, members of the weasel family, were once found throughout North America wherever wetlands existed. Built for a life in the water, they have streamlined cylindrical bodies covered with waterproof fur, short legs, and long tapering tails. Their ears and nostrils can close when submerged. Large males can be three and a half feet long and weigh thirty pounds. Up to five otter pups per litter are born in shoreline dens that are often usurped beaver cavities. An otter's diet consists mainly of fish, which they are remarkably adept at catching. Depending on availability they also consume crawfish, crabs, mussels, snails, rats, and snakes. In Louisiana otters have no natural predators when they are in the water except humans and alligators. When on land they can become prey for bobcats, coyotes, and dogs.

As top predators in their aquatic environments, otters are very susceptible to pollution. The impacts of contaminants in fish are concentrated and multiplied in otters that eat them. Pollutants such as oil remove the protective waterproofing of their fur. The direct loss of essential wetland habitat has contributed to otter declines nationwide, and in some areas trapping for their valuable pelts has been a contributing factor.

During the twentieth century, otter populations in Louisiana plummeted except for those in the coastal region and the Atchafalaya Basin. Otters were extirpated from many inland areas, including Bayou de l'Outre, which suffered saltwater contamination associated with the natural gas industry. In other parts of the country, especially the Midwest, some states lost all of their otters. Ironically, otters are now thriving in many of these same areas as a result of restocking programs that obtained otters from coastal Louisiana. With improved water quality and restrictive trapping regulations, Louisiana otters are once again common throughout the Bayou State, and some have even settled in as Yankees farther north.

Dogwood

One of the most beloved understory trees in southern landscapes is the flowering dogwood, with its bright green leaves and blooms that consist of clusters of tiny inconspicuous flowers surrounded by large, white, petal-like bracts. Native to the eastern half of America, flowering dogwood grows to thirty feet in height and was historically a major component of upland hardwood forests. In Louisiana, it is one of three species of dogwoods native to the state. The others, swamp dogwood and rough-leaf dogwood, have less noticeable flowers and tend to grow on wetter sites.

Flowering dogwood fruits are small olive-like drupes that turn bright red in the autumn. They are relished by many kinds of wildlife, especially bluebirds and other thrushes. Dogwoods in their leaf litter provide an important source of calcium for forest soils. The hard, dense wood of dogwood was used for tool handles, loom shuttles, and the wheels of roller skates. As a medicine it was used as a quinine substitute to treat malaria in humans and as a remedy for mange in dogs, which may be how it got its name.

At the beginning of this essay I mentioned that flowering dogwood *was* a major component of many upland forests. That is decreasingly the case today. In many areas, especially the hill country of Louisiana, all native hardwoods have been replaced by biologically impotent pine plantations with zero tolerance for plant diversity. Those dogwoods that survived the onslaught of tree farms have recently been attacked on another front. An exotic, invasive fungus that causes a disease called dogwood anthracnose is killing dogwoods across the eastern United States. A recent study reported that at least 49 percent of wild dogwoods have succumbed to the disease. Here in north Louisiana it is rampant as cankered trunks are all that remain of many mature trees. Unless there are some disease-resistant trees that survive to pass along the immunity to their offspring, wild dogwoods are on their way to becoming as scarce as hen's teeth.

Coyotes

While motoring up Bayou D'Arbonne recently I rounded a bend to see an animal swimming across the channel ahead. Racing forward I discovered the critter to be a large male coyote. As I pulled alongside he rolled his eyes and yipped a cry of panic. It suddenly occurred to me that desperate wild animals often behave in unpredictable ways and that sharing a small boat with a wet, mad coyote might be an undesirable close encounter. I backed off, he churned forward to the sandbar, and with the long, effortless lope of his species disappeared into the swamp. His is a tale of legends, both prehistoric and modern.

In spite of government efforts to control coyote numbers that resulted in the deaths of hundreds of thousands of coyotes in the last century, the species has expanded its range from the western half of North America and now thrives in every state except Hawaii. In Louisiana coyotes have completely replaced our historical native canid, the red wolf. Indeed, some scientists think that red wolves were hybrids of gray wolf and coyote crosses. Coyotes sometimes mate with domestic dogs with the resulting coydogs having characteristics of both parents. Sightings of very large wild canids are almost always coyote/dog hybrids but nevertheless fuel misconceptions that wolves still exist in Louisiana. The average coyote weighs between fifteen and forty-six pounds.

Unlike gray wolves, coyotes are successful in the modern world because of their adaptability. At home chasing jack rabbits across a vast prairie or munching June bugs under a suburban street light, coyotes are versatile and opportunistic. They eat mostly wild mammals, but fruits and vegetables, wild or otherwise, are significant dietary components when available. Too, coyotes are comfortable having close human neighbors, so comfortable that they are known to snatch a pet dog or cat on occasion. They are not loners like Wile E. Coyote of cartoon fame but often live in a typical pack of six closely related adults, yearlings, and pups.

Within Native American cultures the coyote mythos is rich. He appears variously as a messenger, a fool, or a trickster. Often he has the power of a creator or transformer. Today, his presence in the Bayou State is beyond mythical. He is here to stay.

Bullfrogs

Eighty years ago, wherever there was freshwater habitat in the state of Louisiana, it would have been difficult to sleep during the warm nights of May. The booming territorial choruses of our largest frog, said to resemble the roaring bellows of bulls, were common. Now greatly diminished in numbers throughout their range, bullfrogs are native to the eastern half of the United States.

Very large bullfrogs can weigh almost two pounds and have head/body lengths of eight inches. Females are larger than males. Their breeding season lasts about two months and peaks in May in Louisiana. During this period bullfrogs congregate as females are attracted to the calling males. When a female selects a mate, she

Bullfrogs

lays up to twenty thousand eggs in nearby aquatic vegetation. They are fertilized externally by the male and soon hatch into tadpoles. In northern latitudes the frogs may remain in the tadpole stage for two to three years, but in Louisiana most develop into frogs within a few months. Young frogs grow rapidly, and adults are voracious predators, eating any live animal they can capture and subdue. Insects and crawfish are major food items, but bullfrogs also eat fish, snakes, birds, mice, bats, and other frogs. Bullfrogs are themselves the natural prey of snakes, fish, raccoons, egrets, and herons.

During the first half of the twentieth century, bullfrogs were a valuable commodity in Louisiana. Most towns in the state had agents who purchased frogs for larger companies. Thousands of people hunted frogs at night and were paid up to three dollars per dozen for live frogs. By the late 1920s better restaurants in Chicago, Philadelphia, and Boston were serving Louisiana frogs in a dish labeled frog legs "a la Newberg." The market may have peaked in 1935 when 2.5 million pounds of frog legs were collected statewide. Such a harvest was not sustainable, and frog populations began a decline from which they have never recovered. Since then, bullfrogs have been introduced into many countries around the world and are even considered invasive pests in some areas. Frog legs now served in American restaurants are most likely imported. In Louisiana, overharvest and the loss of critical wetland habitat have insured less raucous spring nights in our swamplands, a deficit not inconsiderable for some of us.

Louisiana Whoopers

Magnificent whooping cranes, once a unique component of Louisiana's diverse fauna, disappeared from within the state's boundaries and almost from the planet. My wife, Amy, shares their current status in the following summary:

Whooping Cranes—Welcome Back to Louisiana!
An extraordinary event happened on April 11, 2016. A wild whoop-

ing crane chick hatched in Louisiana for the first time in more than seventy-five years. The chick's tall white parents, both captive-raised birds, tended the tiny ball of rufous down. Two days later another chick hatched out. Sadly, the second chick died during its first month of life. The first chick's identification tag is LW1-16, and a blood sample confirmed that she is a female.

The whooping crane parents and chicks are a part of an amazing program started in 2011 called "Reintroduction of Whooping Cranes to Southwest Louisiana." This is the remarkable story of the rare and endangered whooping crane and the people who want to save the birds from extinction and reestablish a wild nonmigratory population in Louisiana.

The entire population of whooping cranes almost dwindled away during the twentieth century. In 1942 only twenty-four birds remained. They almost became an extinct species instead of an endangered one. Many factors are to blame, but the major ones were the drastic loss of critical wetland habitat and the lack of effective regulations to protect the birds and their nests from hunters and collectors.

Standing five feet in height, the adult whooping crane is the tallest bird in North America and the rarest of fifteen crane species in the world. Both male and female are covered with a cloak of snowy white feathers and have solid black primary feathers on their wingtips that are visible only during flight. Their heads have a distinctive red patch of skin on top, and they sport black mustaches along the sides of the face behind a long, pointed beak. The majestic birds fly with their black legs outstretched and their neck extended forward. Their wingspan is an incredible seven to eight feet, and the black wingtips spread open like a fan. Their signature call is a loud, bugling, and musical "kar-r-r-o-o-o, kar-r-r-o-o-o" that carries for miles. Their nickname is "whooper." They are stately as they stride on long black legs across the marsh and elegant as they soar above it. With an intriguing social behavior, they mate for life, have distinct vocalizations, and different personalities.

The last whooping cranes to live in the marshes of south Louisiana dwindled away in the mid-1900s. In 1939 biologist John Lynch of the U.S. Bureau of Biological Survey counted thirteen whooping cranes in the marshes north of White Lake in Vermilion Parish, but after a hurricane and flooding in 1940 the biologist could find only six birds. The flock continued to shrink until two birds remained in 1945. In 1947 there was one bird left. A team of biologists captured the last crane in 1950, named him "Mac," and moved him to Texas to live with the migrating flock of whooping cranes. It was hoped that Mac would have a better chance of survival with this flock of thirty-four birds. This larger group of whooping cranes spends the winter on the Texas gulf coast at Aransas National Wildlife Refuge and migrates to northern Canada to summer breeding grounds in Wood Buffalo National Park. The migration is twenty-five hundred miles one-way. Mac, however, did not survive in Texas and was later discovered to have been a female.

Why did whoopers disappear from Louisiana? Two primary reasons surface: unregulated hunting of the cranes, and conversion of their wetland and prairie habitat to farmlands. Whooping cranes are aquatic wading birds and depend on a variety of wetlands to live and raise their young. When wetlands are drained, whoopers have a difficult time finding food and safe places to roost. Destruction of the Louisiana wetlands and prairies, along with hurricanes, took a toll on the last cranes.

Cranes need precious time and suitable habitat to reproduce. Although they can live twenty-five to thirty years in the wild, they do not reach reproductive maturity until they are three to five years old. Their nests are on mounds of marsh plants in freshwater marshes to help avoid predators. Cranes lay one or two eggs each year. After hatching, the four-inch-tall chicks are especially vulnerable in their first three months of life to predation by coyotes, foxes, and bobcats. They have adapted by growing fast (an inch a day), are full sized by three months, and fledge at eighty-one to ninety-one days. Juveniles are dressed in a cloak of copper-colored feathers before they molt into adult plumage.

～

Louisiana Department of Wildlife and Fisheries biologists released 10 juvenile cranes in March 2011. They were hatched and reared at the Patuxent Wildlife Research Center. Other captive-breeding centers have provided chicks in subsequent releases. As of March 2017, 102 cranes have been released in southwest Louisiana, but the project has a long way to go.

Captive-rearing is a labor-intensive process where the young are reared by technicians garbed like adult whooping cranes to keep them from imprinting and bonding to humans. Bonding to anything other than whoopers during the critical chick phase of their life can disrupt future breeding behavior and success. The crane costume consists of a white hood and gown with a long-necked crane puppet head on one arm and a wing on the other. No human sounds are allowed, and the technicians carry recordings of the crane's purring brood call. The chicks learn to eat, drink, and follow these "cranes." In essence they begin to learn how to be a wild whooping crane.

Each year the group of juveniles, called a cohort, is flown to Louisiana in special crates when they are six to seven months old. After a health check and being fitted with leg bands and radio transmitters, they are released into a fenced pen in the marshlands of White Lake Wetlands Conservation Area in Vermilion Parish. Here they acclimate to their new home and learn to eat a Louisiana cuisine including crawfish, tadpoles, frogs, and snakes. Whoopers are omnivorous and adapt quickly to available food sources such as roots, tubers, and berries. After two to four weeks of the cohort's captivity, the gates to the pens are opened and the cranes are free to pursue a life in Louisiana. Some, however, have decided to venture into Texas, and several have flown as far north as Dallas. Fortunately most of these new residents are adapting to altered habitats in Louisiana and finding what they need in rice and crawfish fields. Since whooping cranes do not travel in large flocks, but rather in pairs or small family groups, private landowners are letting them visit, stay, and nest in their fields.

The Louisiana whooping crane population is considered a Non-

Essential Experimental Population (NEP). This special designation within the Endangered Species Act protects the birds yet still allows human activities in the reintroduction area. The population is considered "experimental" because it is being (re)introduced into suitable habitat that is outside of the whooping crane's current range, but within its historic range. It is designated "non-essential" because the survival of the whooping crane as a species would not be reduced if this entire experimental population is lost. The goal of the Louisiana reintroduction project is to establish a self-sustaining population with approximately 120 individuals and thirty productive pairs and to maintain these levels for ten years without the need of additional restocking.

Despite their being an NEP species, it is illegal to harm whooping cranes, for they are protected under applicable state laws and the federal Migratory Bird Treaty Act. Unfortunately, ten of the released birds have been illegally shot and their bodies found. Two young people were apprehended after they shot two cranes of the 2010 cohort. In January 2016, two cranes that had wandered over to a Texas rice field were shot. The offender was caught, sentenced to five years of probation, and ordered to pay nearly $26,000 in restitution. None of the other crimes have been solved.

Under a clear blue Louisiana sky with the wind whispering through the tall marsh grasses the young whoopers of the 2013 cohort strolled and searched for local delicacies. Even the gangly youngsters are graceful. One lifted its wings to show off the new black wingtips and agilely leaped a few feet out of the water. An adult from another cohort circled and called to the new arrivals. Seeing and hearing these birds as they glide across the Louisiana landscape after being gone since the mid-1900s is an experience that I will never forget.

Will these magnificent and charismatic birds find the habitat and security to continue their species in Louisiana? The verdict is still out, but in March 2017 the young female LW1-16 has separated from her parents, donned her adult plumage, and joined a group of older cranes. Her life provides a glimmer of hope for this ambitious project.

Cow Killers

Boys are impressionable creatures. They hone in on pronouncements that combine adventure and danger. Such was my experience many years ago when I was warned by elders to avoid at all costs an insect with the moniker "cow killer." How could such a beast in our midst not be a call to action? Colloquial synonyms for this organism are "cow ant" or "red-velvet ant." It was claimed that the sting was so terrible that it could dispatch a healthy cow. I set out to catch one in a Mason jar.

It was many years after that encounter when I figured out that my early mentors were not first-class entomologists. So-called red-velvet ants are large, colorful insects about three-quarters of an inch long. They are black with dense patches of reddish-orange hair on the thorax and abdomen. In Louisiana they are often seen in late summer running around in open areas, especially those that are sandy. Females lack wings and do indeed possess a potent sting though verified records of cattle mortality have not come to light. Males are similar but have two pairs of wings and cannot sting. And here's a bit of relevant trivia: they are solitary wasps, not ants. They differ from ants in having straight antennae rather than antennae with a jointed "elbow." Ants also have a much narrower "waist."

The life cycle of cow-killer velvet ants is fascinating. Females seek out ground-nesting bees such as bumble bees and lay their eggs inside the nest. The eggs hatch and develop into larvae that feed on the bumble-bee larvae. Other species of velvet ants parasitize certain types of flies and beetles, thus serving as natural control agents. Adult velvet ants feed on nectar.

Velvet ants are not aggressive and will try to escape when confronted. This was the case before I finally trapped a big female under my Mason jar. She was indeed beautiful to look at and much to my surprise emitted a high squeaking sound when I shook the jar. This unusual characteristic of the species and the thrashing whip-like needle of her stinger left an impression that has lasted long beyond my boyhood.

Golden Gar and Other Strange Creatures

Wild critters walk, fly, and swim among us Louisiana folks. While all are interesting, a few are downright strange, or at least strange looking. The unusual appearance of some animals is often caused by skin aberrations usually linked to genetic abnormalities. Albinism is an example. An albino organism cannot produce the dark pigment melanin, and the animal appears white. Many records of albinism have been recorded in species native to our state, including squirrels, raccoons, deer, and various birds and reptiles. There

Gar

are famous albino alligators from Louisiana in zoos around the country. Another similar condition is leucism. A leucistic animal is white because it is incapable of making any color pigments on all or part of its body except the eyes. Thus, leucistic animals will have normally colored eyes in contrast to albinos, whose eyes appear red because internal blood vessels are not masked by the dark pigment melanin. Leucistic animals are rarer than albinos. When eighteen leucistic alligator hatchlings were discovered on a nest near Houma in 1987, it made international news. Melanism can be considered the opposite of albinism in that affected animals produce an excess of melanin and appear black. Melanistic animals are much more common than albinos, probably because they are not as glaringly obvious to predators in their environment. Melanistic fox squirrels are locally common in parts of Louisiana, and in the 1930s scientists from the Chicago Academy of Science determined that a high percentage of the red wolves in the Tensas Swamp at that time were melanistic. Today melanism is frequently observed in our coyotes. The rarest abnormal skin condition is xanthochromism (also called xanthism). It is produced by a reduced degree or absence of melanin along with a greatly increased amount of yellow pigment (chromatophores). The animals often appear yellow or golden. Genetics and in some cases diet are implicated in the cause. Birds are most often noted with this trait, including several species native to Louisiana. Recently a very rare spotted gar with xanthochromism was captured in a local lake and released after photos were taken. This "golden gar" is hard evidence of the strange creatures among us.

Tomatoes in Court

For many southern palates, ambrosia can be defined as a home-grown, vine-ripened, freshly sliced tomato. In their long journey to domestication, tomatoes have made a number of interesting stops around the world, none more so than the U.S. Supreme Court. This particular side trip began in 1883 when the U.S. Congress imposed a 10 percent tax on all imported vegetables. One disgruntled and botanically astute importer challenged the law on the grounds that tomatoes were technically fruits and not vegetables. He was correct according to accepted biological definitions. The justices, though, unanimously leaned in the direction of the common person's vernacular, rejected the botanical truth, and the misconception was perpetuated along with the taxes.

The wild kinfolks of tomatoes grow in Central America and along the western coast of South America. From Peru an ancestor of the tomato may have migrated to Mexico, where it was first domesticated. Aztec recipes using peppers, salt, and tomatoes may have been the original salsa. These first tomatoes were small, cherry-like, and grew on a creeping vine.

Very soon after Cortez's infamous triumphs in Mexico in 1521, tomatoes turned up in Europe. Cultivation quickly became widespread after overcoming a few superstitious speed bumps. Often associated with other poisonous and hallucinogenic members in its nightshade family, tomatoes got a bad rap early on. In German folklore they were tied to werewolves, and the Latin scientific name for tomatoes translates to "edible wolf peach." Tomatoes sailed back to North America with the colonists but, maintaining a shady reputation, were largely considered as ornamentals. Suspicions of the tomatoes' safety were not put to rest until the nineteenth century. It is a good thing. Who would we in Louisiana be without shrimp creole and BLTs?

Black Walnut

Among the greatest ecological calamities of our times in north Louisiana has been the conversion of diverse upland forests in the hill country to pine-plantation monoculture. One hardwood species that was historically common though never abundant is black walnut. Other than a few surviving trees planted in old farmsteads, walnuts have almost vanished from this region. They once grew to over one hundred feet tall and were cherished for their important wood and edible nuts. Still common in parts of the Midwest, walnut trees are so valuable, with individual trees worth thousands of dollars, that timber poaching is often a problem.

Walnut lumber is used to make high-quality furniture, flooring, and coffins. It was once the most sought-after wood for beautiful gunstocks, but has been increasingly replaced by plastic for this use. Ironically, the petroleum that is processed into plastic is often produced using drilling fluids that have ground walnut hulls as a major component. The ground shells are also used in water filters, cosmetics, and abrasives.

Walnut fruits are nuts encased in a hard shell that is embedded in a softer green husk. The nutmeats are difficult to extract but have a unique, natural flavor much richer than that of English walnuts commonly found in grocery stores. Black walnuts are used in gourmet baked goods, ice cream, salads, and pasta dishes. The husks contain chemicals that were once used to make a dark brown dye. When I was a boy, local trappers would boil their steel traps in a tub filled with walnut husks to camouflage the traps and remove human scent.

The ecological role that black walnut played in Louisiana forests in unclear. Certainly, wildlife ate the nuts. Walnut leaves contain chemicals called "polyphenols" that repel insects. Roots produce another chemical that inhibits nearby competing plant growth. The vegetation under a black walnut tree looks different from that found under nearby trees. Whatever the function once provided by the scattered wild, black walnuts in our natural forests, it has been lost in the deserts of genetically modified pine trees that dutifully march in straight rows to the horizon.

Plains Pocket Gopher

As a boy I never looked forward to hay-cutting time. It seemed to be scheduled for the hottest days of summer, and stacking the bales in a low, tin-roofed barn aggravated the situation. Also, I was subjected

Plains Pocket Gopher

to bewildering episodes of informal animal taxonomy. Blistering spears of profanity were sometimes launched by the driver of the hay-cutting tractor, and were triggered by a small, bare-tailed mammal with buck teeth that was derisively called a "damn salamander." For years I was confused about this nomenclature as I couldn't imagine how a tiny, lizard-shaped amphibian could cause such a ruckus. It turns out that the guilty parties were not really salamanders but rather pocket gophers. The colloquial name "salamander" supposedly derived from the phrase "sandy mounders," which is where the trouble begins.

Pocket gophers are small fossorial rodents, which means they burrow and live in underground tunnels. The type in Louisiana is called a plains pocket gopher and is roughly found in the western half of the state. None are found east of the Ouachita River except for a small population on the Bastrop Ridge. They most often live in areas with friable, light-textured soils that suit their burrowing habits. Soils with high clay content or those that are often saturated with water are not good gopher habitat. They are about ten inches long, including the short tail. With short hair, tiny bead-like eyes, heavily clawed forefeet, and muscular shoulders and arms, they are well adapted to a life of tunneling. Two huge rat-like incisors and fur-lined cheek pouches that open to the outside on each jaw support their exclusively vegetarian diet of roots, bulbs, and tubers of various plants.

Pocket gophers should not be confused with moles, which are much smaller and leave different telltale signs. Moles burrow also, but their tunnels are much shallower than gophers', and the tops of mole tunnels often protrude above and snake along the surface of the ground. The marks of gophers are fan-shaped mounds of earth that somewhat resemble fire-ant hills. It was these sandy mounds of the "salamanders" that sheared the hay-mow bolts and set off the expletives in my bygone hay-cutting days.

Gray Fox

When people experience intense emotions such as fright or awe, they often remark that they feel their hair standing on end. Startling night sounds that emit from our local forestlands are sometimes a source of these involuntary chill bumps. Owls, especially barred owls with their wild screams and hoots, have sent many a novice outdoorsman packing. There is one species of local mammal, though, that can hold his own with owls when it comes to nocturnal caterwauling. At my house in the woods they begin tuning up mostly during their winter breeding season. A sudden explosive burst of harsh, coughing barks evokes images of mythological hellhounds and this coming from the vicinity of my compost pile that I often visit after dark. The origin of this racket is known by its Latin scientific name as *Urocyon cinereoargenteus,* which translates to silvery, gray-tailed dog. We know them as gray foxes. It's hard to believe that such a volume of noise can be produced by an animal weighing an average of fourteen pounds. Gray foxes are found throughout Louisiana, except the coastal marshes, and prefer upland habitats of mixed pine/hardwood forests. They are primarily carnivorous and relish rats, mice, rabbits, and insects but also consume acorns, berries and other wild fruits when available. The slightly larger red fox is a close cousin and is sometimes found in the same habitat. Both species are esteemed quarry in the declining sport of fox hunting. Unlike red foxes, grays can climb trees to forage and escape predators. At one time most people in this area who lived in the country raised poultry, to the delight of the gray fox. No longer a major nemesis, the gray fox has a connection to poultry today only via the goosebumps caused by his otherworldly clamor.

Gourd Heads

Soaring gracefully overhead with a wingspan exceeding five feet, wood storks are more attractive at that distance than when up close in person. With snow-white plumage, except for a black tail and trailing wing edges, they are the only true stork found in North America. It's their naked gray head and neck that only a mother wood stork could love. Add a large, thick, slightly curved bill, and the common name "gourd head" is not totally inappropriate.

Although wood storks bred in Louisiana until as late as 1918, those that are observed in the state now are from south of the border. In late summer they begin appearing in cypress swamps and flooded fields. Their arrival is always timed to occur as these wet-

Gourd Head

lands are drying up. The shrinking pools harbor a banquet of easily captured stork foods such as small fish, frogs, crawfish, and large insects.

Two distinct populations of wood storks exist. One group nests in Florida in large colonial clusters often high in the tops of trees. After breeding, these birds disperse to spend the summer along the East Coast as far north as the Carolinas. Florida's dwindling wetlands, especially in the Everglades, resulted in perilously low populations and Endangered Species status for the remaining birds. Conservation efforts helped, and the eastern wood storks were upgraded to Threatened status only recently.

The Louisiana wood storks have different roots. For many years, one of the last remaining mysteries of North American bird migration was where our Louisiana storks originated. The puzzle was solved when several of the birds were captured and fitted with satellite-monitored radio transmitters. Upon departing Louisiana in the early fall, our storks fly south to Mexico and Guatemala, where they breed in coastal estuaries. Come next summer, though, something mystical will occur inside their "gourd heads," and they will head north to Bayou Country.

Red Wolf Revelation

John James Audubon's son, John Woodhouse Audubon, was a pretty good artist in his own right. When the elder Audubon began showing signs of mental illness toward the end of his career, his son stepped in to complete their famous portfolio on American mammals. The topic of this essay is the younger Audubon's Plate No. 82 from the year 1845, which depicts a wolf standing on a sandbar in a Texas river, sniffing a bison horn surrounded by scattered mussel shells. He labels it the "Red Texan Wolf." The status of this animal is headline news today, at least in the biological realm.

Weighing forty-five to eighty pounds, red wolves are larger than coyotes, with broader muzzles, but much smaller than gray wolves of the American West. Their name derives from the reddish color of the fur behind their ears and on their neck and legs. However, in the early 1930s researchers from the Chicago Academy of Sciences discovered that many in northeast Louisiana were melanistic, being totally black. Socially, red wolves live in packs of five to eight family members. Breeding pairs bond for life and have one litter per year. Their diet consists of small mammals such as rabbits, raccoons, rodents, and occasionally deer.

Red wolves were the large canid that once inhabited most of the southeastern United States, including all of Louisiana. Like wolves everywhere they were ruthlessly pursued in predator-control programs until their populations were decimated. By the 1960s the only remaining red wolves lingered in a small area of prairie along the Texas and Louisiana coast. By this time they were considered a distinct species, *Canis rufus,* and in 1973 were listed for protection under the Federal Endangered Species Act. Seventeen of the remaining wolves were captured to begin a captive breeding program, and the species was declared extinct in the wild in 1980. Offspring from the captive animals were released on a North Carolina refuge in 1987 and later in the Great Smoky Mountain National Park. The release failed in the Smokies, but about seventy-five red wolves roam their historical range in eastern North Carolina today. Another two hundred are in captivity at several places.

These extensive recovery efforts have been employed based on the premise that the red wolf is a distinct species. Recent molecular research on the DNA of North American wolves and coyotes in today's news has stirred the scientific pot; indeed it has revealed an unexpected gumbo. The large genome study found that there is only one species of wolf on the continent—the gray wolf. Red wolves are actually hybrids with genomes that are 75 percent coyote and only 25 percent wolf. The findings complicate the legal and political world as the Endangered Species Act has no guidelines that address hybrid animals, and I'm wondering if the yips and yaps of coyotes behind my house are their hilarious reactions to the muddle. Even the Audubon boys wrote that their Texas Red Wolf was nothing more than a color phase.

Periodical Cicadas

A loud, piercing, unrelenting, high-pitched whine that permeates my auditory nerves is how I would describe the noise. For the last few days it has been constant in parts of Louisiana and across the South. It began abruptly, will persist for several weeks, and then disappear completely. The noise will not be heard here again until the year 2024. Once the subject of a scientific mystery, periodical cicadas are the source of the sound. Cicadas mind you, not locusts; that's a different bug. Periodical cicadas are harmless insects that in our area have thirteen-year life cycles. Those in northern states tend to have seventeen-year life cycles. Both are different from the familiar annual cicadas that drone away the late summer evenings every year.

The current brood has spent the last thirteen years underground as nymphs sipping on root juices. Then, as if by magic but actually tied to soil temperature, the developmentally synchronized young emerged all at once. A tell-tale sign of their emergence is an abundance of holes in the ground about the diameter of a man's finger. They soon crawl up a tree, shed their skins, and harden into red-eyed adults. In some instances there may be more than 1.5 million individuals per acre. Even though only the males call to attract

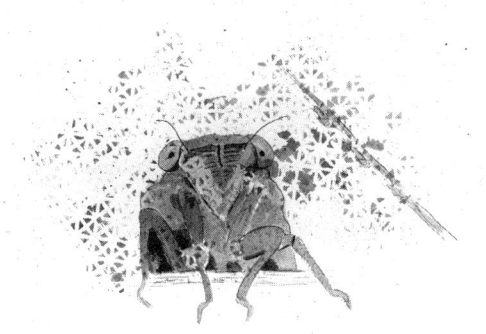

Periodical Cicada

mates, that's a lot of noisemakers. After mating, females slit plant
stems with their ovipositor and lay eggs inside to renew the cycle.

Plants rarely suffer long-term damage from the insects and,
for some animals, a periodical cicada emergence is a time of ser-
endipitous feasting. Many types of birds from brown thrashers to
wild turkeys dine on them. Likewise, mammals such as opossums,
raccoons, and foxes take advantage of this fleeting source of high
protein. Even fish gorge on cicadas when thousands fall into bayou
waters. For them the banshee wails sound like dinner bells.

Palmetto I

During the Civil War, southern white women were often deprived
of store-bought goods. Hard times required that they adopt the cre-
ative strategies of hapless slaves for basic necessities. Manufactured
hats in particular were scarce as all production was geared to outfit
southern soldiers constantly exposed to the elements. As a result,
domestic millineries cropped up in countless households. The most
common raw product used to produce thousands of hats was a type
of palm that we call palmetto.

Along the eastern seaboard, cabbage palm is the dominant palmetto. Its fan-shaped leaves grow on tall straight trunks to eighty
feet in height. Inland from the Gulf Coast, including all of Louisiana, dwarf palmetto is most common. In this species only the
characteristic leaves are aboveground, the stem being buried. It is
usually less than six feet in height. Deer will browse the tender new
fronds, and the attractive pale flowers yield hard black fruits eaten
by birds and raccoons.

Landscapes with dense stands of palmetto are memorable. A
refugee traveling through the Boeuf River Swamp of northeastern
Louisiana in 1863 wrote in her diary: "For a mile the road was a
beautiful avenue through this forest, then immediately the character of the scene changed, the large beautiful trees were still there,
but around their roots the palmetto grew thick, one who has never
seen it can have no conception of the effect, the scene was tropi-

cal indeed, from the forest we emerged into an open space covered thick with glossy dark green fans of palmetto."

Another woman in Morehouse Parish described the local situation: "The seclusion and inaccessibility of the place made it difficult to obtain very elaborate wearing apparel. Palmetto grew abundantly and luxuriantly around our home, and we became experts in weaving it into hats which were very pretty and unique. The palmetto was gathered and then boiled. The boiling process bleached it perfectly white, and made it soft and pliable, thus adapted to the use we made of it." Soldiers, too, used the palmetto plant for a variety of purposes from building material to screens for camp latrines. None were more inventive than hat-making by the women back home.

Palmetto II

Have you ever noticed that, like hair styles, clothing, and home décor, landscaping plants are trendy? For those in the nursery business the development of a new variety of rose or azalea is just as coveted as a fashion designer's spring wardrobe. Sometimes a new plant variety actually has long-term merit as in the case of a disease-resistant iris or a cold-tolerant hibiscus. In many instances, though, new plants are only passing fads or outright bad ideas. Consider the recent trend to plant palm trees in areas far outside their natural range, including north Louisiana. Someone sold a faulty bill of goods to a lot of folks that resulted in the recent winter kill of hundreds if not thousands of palms across the region. There is a reason that those types of palm don't grow here naturally, and in spite of ongoing climate change it will probably be a while before they thrive locally.

In many cases there already exists a native plant similar to the latest fad. Dwarf palmetto is a native palm that grows throughout the swamps of the southeastern states. Abundant in our local wetlands, it makes an excellent landscape plant. As a type of fan palm, the plant has glossy green leaves that may be four feet wide. Best of all, it is adapted to our climate and grows well with virtually no pampering necessary. So there you have it, another good reason to ignore the passing vogue and use native plants in your landscape at every opportunity.

Tiger Owls

At the very mouth of the Mississippi River there is a small island that once served as the headquarters of Delta National Wildlife Refuge. A surplus fire tower was erected on the site in order that the wardens might watch for poachers in the vast flounder-flat marshes of the delta. A friend who worked there once told me that for several years the tower was deemed unsafe and off-limits for a couple of months each winter. It wasn't because of high winds or lightning storms that the hundred-foot tower was condemned but rather the presence of birds that some people called Tiger Owls. This back-

Great Horned Owl

woods moniker was attached to great horned owls by people knowledgeable of their innate fierceness. Great horned owls are found across most of North and Central America and a large part of South America. Indeed in many areas they are the apex predators of the skies.

Great horned owls are found throughout Louisiana but are not as common as the smaller barred owls and diminutive screech owls. They can stand almost two feet tall with a wingspan greater than four feet. Mottled and striped in brown streaks and with large yellow eyes, they are named for two prominent ear tufts. Their calls are low, haunting hoots that mean business when they are defending territory.

Great horned owls use keen senses of sight and hearing to hunt their prey—almost always at night. Formidable talons coupled with modified flight feathers for silent approach make them stealth predators of darkness. Their list of prey includes rabbits, squirrels, skunks, armadillos, reptiles, other birds, and even young foxes and coyotes. In many places they were once considered outlaw birds because of their habit of occasionally killing poultry. The fact that they also consumed untold numbers of destructive rodents was not considered. Old Louisiana hunting regulations stated that they "may be killed at any time." Now they are rigidly protected by federal and state laws. Adults generally have no natural predators, and most mortality is still human-related. Owls are killed by collisions with cars, buildings, and power lines. Less frequently they are still poisoned, trapped, and shot.

In Louisiana, nesting begins in December or January, among the earliest of all birds. They do not build nests per se but rather use abandoned hawk, crow, or squirrel nests to lay their two or three eggs. When suitable nests of other species are not available to confiscate, they will use large tree cavities or, as in this case, a fire tower in the marsh. At this critical time in their life cycle they are most fierce and will defend their nest "like tigers," even from well-intentioned game wardens.

Box Turtles

That an old, time-marred box turtle in my hand today could be the same one held by my great-grandfather on the edge of this swamp a hundred years ago infers a connection mystical if not spiritual. Though unlikely, it is possible.

The most common land turtle found throughout Louisiana is the three-toed box turtle, a subspecies of the eastern box turtle, so named because it usually has three toes on each hind foot. A second species, the ornate box turtle, is very rare and occurs only in the extreme southwestern part of the state. Box turtles have hinged bottom shells (the plastron) that can close tightly against the upper shell (the carapace) for protection. Their high-domed shells are about five inches long when mature. Males often have a concave plastron, red eyes, and orange markings on the head and front legs. Females have yellow or dark eyes, duller markings, and a flat plastron.

Most breeding and egg-laying occurs in the summer. Females dig a hole and lay three to eight elliptical, thin-shelled eggs that hatch in about three months. Box turtle eggs and hatchlings suffer high mortality rates. It takes five to seven years for the young turtles

Eastern Box Turtle

to become sexually mature. They are omnivorous and eat a variety of plants, insects, and other animals including flowers, roots, berries, mushrooms, earthworms, snails, slugs, beetles, and caterpillars. They survive cold winter temperatures by burrowing into the leaf litter and becoming dormant. Their wintering site is called a hibernaculum.

Although they may be found in a variety of habitats, three-toed box turtles are primarily a woodland species. Adults have a home range of two to five acres and exhibit high site fidelity, meaning that they don't roam very far in their lives. If moved by humans, they try to return home, an act that often results in their deaths on roads. Other threats include loss or fragmentation of habitat due to development, unnatural fire regimes, and collection for the pet trade. Because of their delayed sexual maturity, low reproductive rate, and high mortality in eggs and young turtles, the loss of a very few adults can cause a population to crash in any given area. However, for those that avoid the hazards, natural and otherwise, they have the innate capacity to live more than a century. I like to imagine the old-timer crossing my driveway and nibbling the mayapple fruits of a late spring morning was also the youngster noticed by my great-grandfather on his morning amble to the shallow well to draw a pail of kitchen water a hundred years past.

Pallid Sturgeon

For a very long time, seventy million years or so, a strange sort of fish has been swimming along the sandy bottoms of North America's largest rivers. They have neither bones nor scales. Instead they have a cartilaginous skeleton and rows of boney scutes for protection. An elongated snout, an asymmetrical, scimitar-shaped tail, and whisker-like barbels add to their bizarre appearance and reputation as living fossils. More people are familiar with their eggs than with the fish themselves. Eight species of sturgeons inhabit this continent, and three are found in Louisiana, if only barely. They are among the rarest of our aquatic fauna.

Of the three types of sturgeons in Louisiana, perhaps more is known about pallid sturgeons due to a recent flurry of research after it was placed on the federal Endangered Species list in 1990. Named for its pale coloration, pallid sturgeons are found in the Missouri and lower Mississippi river basins. In Louisiana they inhabit the Mississippi, Atchafalaya, and Red rivers. Attaining five feet in length and weighing up to eighty-five pounds, pallid sturgeons are long-lived but slow growing, reaching maturity only after fifteen years. They are opportunistic bottom-feeders of invertebrates and small fish.

It is doubtless that pallid sturgeons were once more common in suitable habitat throughout their range. In Louisiana and elsewhere they were sought commercially for their flesh. One record indicates that in 1914 a single fisherman had 150,000 pounds of sturgeon iced down at Melville on the Atchafalaya River. However, they were most prized for their eggs, from which valuable caviar is made.

Although commercial fishing may have reduced local stocks, levee building and lock construction on major rivers decimated the species everywhere by reducing turbidity levels and flow velocities beyond those parameters necessary for its survival. For decades, biologists observed no natural reproduction in the species, and the remaining fish seemed to be aging out. In Louisiana at least, pallid sturgeons are hanging on, but barely so. As it isn't likely the forces that caused the decline will be reversed anytime soon, the prognosis for this species is not good even after a showing that has lasted seventy million years. Such is the heart-rending dominance of our own species.

Whirligig Beetles

A boyhood on the edge of a Louisiana swamp is fraught with danger, some real but most imagined. An example of the latter occurred when as adolescents my neighborhood gang would gather at the White's Ferry Bridge to swim on hot, summer days. The event began as we jumped from the high bridge into Bayou D'Arbonne below.

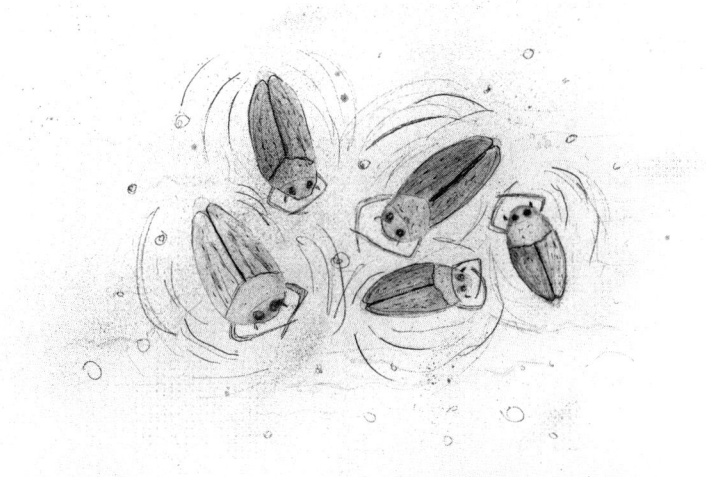

Whirligig Beetles

The older boys always warned us of an instant death that would befall us should we be so unfortunate as to do a belly-buster from that height. Next in the line of perils was the alleged near-death experience of jumping into a swarm of whirligig water bugs that frequented the placid water of the bayou. Doubtless, they would be forced up into one's cutoff jeans where in such an agitated state they would bite and sting any and all available, sensitive flesh. We thought about these things a good bit.

The whirligigs in Bayou D'Arbonne and most Louisiana water bodies are a type of beetle of which there are more than seven hundred species worldwide. They are often found in large groups, and their name derives from their habit of swimming rapidly in circles when alarmed. Most are less than a half-inch long and lustrous black with divided eyes that are believed to enable them to see above and below the surface. The front pair of legs are long and slim, while the middle and hind pairs are short and flattened to function as paddles. To escape danger they dive under water, carrying a bubble of air attached to the tip of the abdomen to allow breathing.

Surprisingly, most species have well-developed wings and can fly. They lay their eggs under water attached to vegetation. Eggs hatch into predaceous larvae. The adults also are active predators on other pond insects. As such, they play a role in the balance of nature and are considered very beneficial to have around. However, as they neither bite nor sting humans, the part they play in keeping juvenile boys in check is an artifact of vivid imaginations.

Pawpaws

During one of the earliest European explorations of interior North America, in 1541, Hernando de Soto's scribes wrote of a peculiar tropical-like fruit that was being cultivated by Native Americans. When President Thomas Jefferson sent William Dunbar and George Hunter to explore the Ouachita River in 1804, Hunter recorded a small bayou named after this plant that entered the river on the east side about a league above the mouth of Bayou Bartholomew. Two years later, as the Lewis and Clark expedition neared the end of their epic journey to the Pacific Northwest, they depended on the fruit of this plant for sustenance when their rations ran low and game was scarce. Many of us still sing a jingle about picking up the fruits and putting them in our pocket.

The plant in all these scenarios is pawpaw (*Asimina triloba*), a small understory tree found throughout most of Louisiana except for the coastal parishes. Rarely more than thirty-five feet tall, it grows best in fertile, well-drained soils. Dense, clonal patches form as stems sprout from underground rhizomes. Small maroon flowers appear in spring before the new leaves and have a very faint odor of rotting meat. Not surprisingly, they are pollinated by the likes of blowflies and carrion beetles. When I once mentioned to a seasoned botany professor that my young pawpaws were not producing fruit, he suggested tying a piece of meat in the tree to attract the pollinators.

It is these fruits that have attracted the attention of humans for thousands of years. They are considered the largest edible native fruit in North America. Yellowish-green, they are two to six inches long and filled with bright yellow pulp and large brown seeds. Their flavor has been described as custard-like, similar to a combination of banana, cantaloupe, and mango. When they are eaten raw or as substitutes for bananas in baked desserts, their nutritional value as measured in vitamins, minerals, amino acids, and calories is greater than that of apples and peaches.

The large pawpaw just off my front porch, though a prolific producer of fruit, doesn't contribute much to my larder. Gray foxes, raccoons, squirrels, and opossums are the usual beneficiaries. And before this tree matured we never saw zebra swallowtails here. Their larvae feed exclusively on young pawpaw leaves. Chemicals in the leaves when ingested make the butterflies unpalatable to predaceous birds. So, history, spectacular butterflies and other wildlife, with a unique healthy dessert to boot—what other reason does one need to have a pawpaw in the landscape?

Emily Dickinson's Robins

Emily Dickinson wrote a poem titled "I Dreaded That First Robin So." The reason that she did not look forward to seeing this harbinger of spring return to her Massachusetts home is often debated. Some say the poem has sexual overtones or expresses her struggle with depression. Had Ms. Dickinson lived in Louisiana, she would have been forced to choose another species of bird as a metaphor because robins don't return here in the spring. They are year-round residents. Found throughout North America, those in the northern part of its range do migrate south for the winter. At this time our local breeding population of robins is augmented by the northern birds for only a season.

The American robin is a songbird in the thrush family. They are familiar to everyone as they hop across lawns, stopping to listen for earthworms with cocked heads in a characteristic manner. About half of their diet consists of worms, insects, and other invertebrates with the remainder made up of fruits and berries. Robins are among

Robin

the first birds to nest in the spring and will often raise two or three broods each year. They have a complex repertoire of songs used to denote territory and warn of predators. Crows, hawks, rat snakes, and especially cats are their enemies. In Louisiana, robins were once hunted for food even though they are protected by federal law. They are also known to carry West Nile virus.

Another analysis of Emily Dickinson's poem, though mundane, might be closer to reality. As a hay-fever sufferer, she may have dreaded the spring arrival of robins simply because it signaled the beginning of allergy season. Regardless of Dickinson's actual motive, the robins are innocent.

Extremophiles

Perspective matters. This is especially true when humans consider the lives of other animals and plants. One group of organisms exists so far out on the edge of our reality that we call them extremophiles.

Extremophiles are usually microorganisms that live in extreme environments that are hostile to most life on earth, including ours. They actually thrive in physical and chemical surroundings that are deadly to others, such as the near-boiling water of deep-sea, hydrothermal vents, or acid-laden rocks within volcanos. Because of our biased perception of where life should exist, the discovery of most known extremophiles didn't begin until the 1980s and 1990s when scientists started to look in unusual places. Now that researchers have determined the wide range of conditions under which life is possible, they have begun to study the remarkable creatures themselves. Already it has been determined that they hold great promise for genetically based medications and industrial chemicals and processes.

This story has an unlikely connection to Louisiana through the beautiful Ouachita River that highlights our natural environment. In the early 1800s, Thomas Jefferson sent out four expeditions of discovery to report on the flora, fauna, geology, and people of the lands in the newly acquired Louisiana Purchase. The most famous of these was the Lewis and Clark Expedition to the Pacific Northwest. However, at the same time Jefferson chose William Dunbar and George Hunter to explore the Ouachita River Valley. From the mouth of the river their crew rowed upstream to Monroe, where they traded for a smaller boat and acquired a guide to lead them on to the rumored headwaters at what is now Hot Springs, Arkansas. When they finally arrived, they set up camp and with their primitive scientific instruments began to study the mysterious springs of scalding water that burst forth from the mountainsides in the narrow valley. On December 28, 1804, William Dunbar put his crude microscope aside and wrote in his journal: "I have always thought it probable that minute animalcules might be found in this water and have always looked attentively to that object; at length I found this evening upon the green matter a minute shell animal, shaped like a muscle or kidney, it is about the size of the smallest grain of sand."

Dunbar had just described one of the first extremophiles, a tiny crustacean that lives in the 145 degree waters of the hot springs. A significant detail to remember on this matter is that the creatures are extreme based on our perspective. From their point of view, it's a perfectly normal world.

Carolina Wrens

Though mates for life, for much of the year they sleep on opposite sides of our house in the woods. One we call the east wren. This is the male. The west wren is the female that sometimes roosts above the front door or in a wind chime that she often rings on a dead calm evening seemingly for her own amusement. They are Carolina wrens, the most common of five species of wrens in Louisiana. Except for marsh wrens that nest along the coast, the others only spend the winters here. A pair of Carolina wrens will establish a territory and live their entire lives in one small area. They are small, brown, rotund birds with a long, downcurved bill and a long tail that often bobs over their backs. A slender white stripe above each eye is characteristic.

They are busybodies and, in addition, the male is a loudmouth. They creep and probe around under my front porch and in the garage, explore the firewood stacks, and if their business is questioned dive headlong into brush piles that I leave for cover in the yard. The east wren defends his territory all year by singing from a large repertoire of loud, whistling calls in a constant, playback loop. Some imagine that at times he is saying "teakettle, teakettle, teakettle." Only the males sing loudly. Females are fussy with a vocabulary limited to stridulous buzzes, chirrs, and half-hearted songs.

When it comes to nesting, they ratchet up the drama. Nest sites can be natural cavities in trees or unnatural coffee cans, tractor radiators, flower pots, or most recently the pocket of my rain jacket hanging in the garage. Apparently indecisive, they sometimes build a half-dozen nests before deciding on the one that is just right. It's a bulky mass of leaves, grass, hair, feathers, and often a shed snakeskin. The female lays four to eight eggs that she incubates for about two weeks. Both parents feed the nestlings for two more weeks until they fledge. In Louisiana, three broods a year are common. Throughout the year their diet is mostly a wide variety of insects and spiders. In winter they occasionally consume wild fruits and seeds and will eat suet from feeders. As neighbors go, they are a never-ending source of entertainment, even if a bit nosey.

Wax Myrtle

High electricity bills are aggravating, but we often forget that in the two hundred thousand years of modern human history we have only been able to complain about this particular problem for less than a hundred years. Choosing as a point in time the early colonial days of Louisiana we note that there were no electric or gas lights, no kerosene lanterns; even dependable matches had yet to be invented. Candles were the main source of lighting, and they weren't cheap. Made from the rendered fat of whales, hogs, and bears, can-

Wax Myrtle

dles for most people were expensive, luxury items. One did not need to be reminded to turn out the lights when leaving a room.

Even then, more economical sources of energy were in demand, and early settlers learned of one in the form of a native plant that we now call southern bayberry or wax myrtle. This broadleaf shrub that grows to twenty feet tall is common throughout the state. Pale blue berries grow on the female plants and are coated with wax. Boiling the berries causes the wax to separate so that it can be skimmed off and molded into candles, and many settlers soon learned to cultivate the plant for that purpose. In 1727 a shipload of Jesuit priests arrived from France and settled on a tract of land just upstream of New Orleans. There they established a plantation of wax myrtle as a commercial enterprise. Soon their product was being exported to other parts of America, as well as France.

As long as you pay your utility bills there is no reason to grow wax myrtle for an energy source today. There are, however, plenty of good reasons to add it to your landscape. It is an attractive evergreen plant with aromatic leaves. It can be highlighted as the focal point of a garden or trained into a hedge. Roots with nitrogen-fixing bacteria allow it to grow well in harsh soils. The energy-laden berries are eaten by some birds, including myrtle warblers, which derived their name from this behavior. The shrub is also the larval host for two types of hairstreak butterflies. Most importantly, it is a native plant, unlike the red-tip photinia and other mass-produced exotics that haunt many yards, and which can't hold a candle to wax myrtle.

Pelicans

In the autumn of 1869 a Monroe newspaper found the following natural history account so interesting as to be newsworthy: "Mr. John Wentzell, living two miles below town, killed a pelican opposite his place on Wednesday, which proved to be one of the largest birds of which we have ever noticed on any account. It measured eight feet, lacking one inch, from tip to tip of the wings, and carried a haversack (forming a part of the mouth) which we judged would hold a gallon of water. Its feathers were pure white, except that its wings were tipped with black, and were very soft and fine."

Indeed, the American white pelican is a large bird and still may be seen migrating through north Louisiana on bright autumn days. In fact, white pelicans are much larger than eagles with a wingspan of almost ten feet that is exceeded only by California condors in North America. The "haversack" of the newspaper article refers to the giant bill and its extendable pouch.

White pelicans breed on inland lakes and marshes of the upper Midwest and migrate to coastal wintering grounds. Large groups can often be seen drifting south in slow spirals as they take advantage of rising currents of warm air. In this region they often stop for several days on the larger lakes such as D'Arbonne, Claiborne, and Poverty Point. Most eventually wing their way to the Gulf.

White pelicans should not be confused with much smaller brown pelicans, which are found only along the coast. The brown pelican is the bird of the Louisiana state flag and seal. They differ in habits as well as appearance. For example, brown pelicans feed by plunge-diving into the water to capture prey while white pelicans make shallow surface dives or dips with their huge bills. Both species eat mainly fish. It would be very unusual to see a brown pelican in north Louisiana.

Both species suffered drastic declines in the 1960s as they bioaccumulated the pesticide DDT in their tissues. After DDT was banned, the birds recovered dramatically and are now common throughout their range. It would not be unusual at all to see white pelicans from John Wentzell's old homeplace two miles below Monroe. It would, however, be illegal to kill one.

Sassafras

In the midst of the Civil War, Kate Stone, a fierce advocate of the southern cause, wrote from a plantation near Tallulah, "The plums and sassafras are in full bloom and the whole yard is fragrant. We

Sassafras

all drank sassafras tea for awhile but soon got tired of it, pretty and pink as it is." At the same time, the infamous Yankee General Benjamin Butler was enjoying the delights of genuine New Orleans gumbos during his occupation of that city. His meals were surely spiced with dried, powdered sassafras leaves known as filé.

Sassafras is usually a shrub or small tree but can grow to eighty feet tall and three feet in diameter in optimum conditions. It often forms dense, shrubby thickets. The deciduous leaves are unusual in that three different shapes may grow on the same plant. Sassafras is widely distributed throughout the eastern and southern United States.

As a medicinal plant, sassafras is reported to be one of the first exported to Europe from the American colonies. Tea brewed from the roots was used to treat fever, pneumonia, bronchitis, catarrhs, measles, and mumps. In recent years safrole, an oil found in the plant, has been reported to be carcinogenic in lab animals.

The wood of sassafras is very durable yet somewhat brittle. It was used for ox yokes, cooperage, light boats, poles, posts, and crossties.

Bedsteads and roost poles in chicken houses were once made of sassafras to deter insect pests. A yellow to orange dye was made from the roots. Although an additive is manmade today, the odor of root-beer drinks once derived from sassafras roots.

Like a host of other plants wild and cultivated, sassafras was used to brew an alcoholic drink during the Civil War. If you are interested, here's a period recipe: "Take eight bottles of [sassafras] water, one quart of molasses, one pint of yeast, one tablespoonful of ginger, one and a half tablespoonful of cream of tartar, these ingredients being well stirred and mixed in an open vessel; after standing twenty-four hours the beer may be bottled, and used immediately."

Enjoy!

Beavers

European exploration of Canada and much of America was driven by a quest to satisfy the vanity of highbrow men in London and Paris. It didn't seem to occur to these folks that their egos were being stoked literally on the backs of the second largest rodent in the world. Not unlike the women's fashion trend for plumed hats

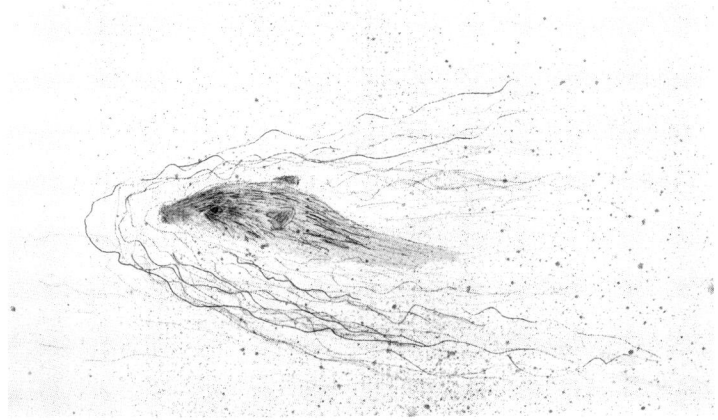

Beaver

that almost drove our wading birds to extinction in the early twentieth century, the demand for beaver-felt top hats extirpated beavers throughout most of their range in North America. The pursuit of beavers for their naturally waterproof hides propelled the earlier exploration of the continent.

Beavers are keystone species in many ecosystems. Legendary dam builders, they construct wetland habitat that would not otherwise exist in some areas. With front teeth that never stop growing, they fell trees and plow up mud to engineer barriers across streams, innately selecting the most suitable site for the job. In doing so, they provide homes for a host of wetland-dependent animals such as trout, waterfowl, moose, and otter.

Beavers use their ponds to access food, provide protection from predators such as coyotes, and raise their young in lodges built of sticks. Each pond usually supports a colony of six beavers that includes a pair of adults, two yearlings, and two young-of-the-year kits. They aggressively defend their territory from other beavers.

Historically blessed with an abundance of wetlands, Louisiana was never as reliant on beavers for wetland habitat as many regions. Nevertheless, beavers played an important role in the ecology of our inland swamps, especially in times of drought. When most of the wetlands in the lower Mississippi Valley were converted to agriculture, human/beaver conflicts escalated as competition for the remaining forested wetlands pitted beaver survival against timber considered commercially valuable. Skirmishes also erupt when beavers encounter the landscaped trees of lakefront homeowners. Egos aside, in many areas today the disputes end in stalemate, and beaver populations persist.

Rain Crows

One definition of the word "lurk" is to lie in wait in a place of concealment. Among those birds that spend time along Louisiana bayous, one species in particular can be said to exhibit this behavior as a matter of habit. Rain crows, often heard but less often seen, are bona fide lurkers as they perch with hunched shoulders that belie a long, graceful neck in a pose that for all the world appears to me an expression of guilt. Of course, my scientist brethren will justifiably gnash their teeth at this anthropocentric characterization of a wild animal and label me a heretic, so I'll take it back, mostly. The part that I am holding fast to is the moniker "rain crow." You won't find rain crows in your field guide or on the checklist of Louisiana birds. Instead, their photo will be near the roadrunner, to whom they are kin, and it will be labeled as the yellow-billed cuckoo. So rain crows and cuckoos are one and the same. They are long, slim birds with swept-back wings and a long tail. Brown above and white below, they have a yellow bill as the name implies and a dark face mask.

Rain crows are long-distance migrants arriving in Louisiana in late spring and departing in early fall. Most spend the winter in South America, often as far away as Brazil and Argentina. With little time to spare along the bayous, rain crows have a very short nesting cycle compared to other birds. Young birds are fully feathered and capable of leaving the nest only seventeen days after the beginning of incubation. Rain crows in America (which include the black-billed cuckoo—a species that only passes through Louisiana) occasionally lay their eggs in other birds' nests but not to the extent of their infamous cousin, the common cuckoo of Eurasia.

The diet of rain crows is unique in that it consists mostly of caterpillars. They are among only a few bird species that eat hairy caterpillars and are especially fond of web-building, tent caterpillars of which they can eat as many as a hundred in one occasion. As natural biological controls for insects that are capable of defoliating entire forests, it is expressly alarming that the continental population of this species has declined 54 percent since 1970 for reasons that are not clearly understood. The ubiquitous, systemic use of pesticides is one suspect.

The call of rain crows is a series of clicking croaks followed by loud coos as from a mourning dove on steroids. They tend to call on hot, humid summer afternoons and respond to loud noises such as the rumble of thunder, a behavior that led to their colloquial name of "rain crow." This name was imprinted on my cultural psyche at an early age by an elderly neighbor who owned a bait stand and cussed the birds for stealing the valuable catalpa worms from his trees. You can call them yellow-billed cuckoos if you wish, but I'm sticking with rain crow as a matter of habit.

Paddlefish

Seemingly unrelated political decisions often affect wildlife resources and lead to a cascade of unanticipated events in the most unlikely of places. That U.S. policy in Iraq and Iran could suddenly influence my daily biological work in the spectacular Lacassine marshes of southwestern Louisiana is a good example. The nexus involves humans' odd obsession with fish eggs, and the story line goes like this. Caviar, the peculiar salt-cured eggs of sturgeon, with origins in the Middle East, was a hot commodity in American markets until the late 1970s when all imports from Iraq and Iran were banned for the countries' alleged bad behavior. Not to be outdone with the loss of two of the world's largest caviar exporters, U.S. markets turned to eggs of native paddlefish as a substitute and prices for this heretofore inconsequential commodity soared. Suddenly, in 1984 the bayous and especially the Mermentau River that braided the Lacassine marshes were filled with nets of nonresident commercial fishermen seeking the abundant and totally unregulated paddlefish resource.

Paddlefish are prehistoric in appearance and physiology and have been around since fifty million years before the first dinosaurs showed up on the scene. They are unique in having a large paddle-shaped snout, called a rostrum, which comprises a third of their body length. Once thought to be used as a shovel for rooting up food items, the rostrum is now known to be filled with electrical receptors that can detect masses of tiny zooplankton, their primary food. As filter feeders, they swim through the water column with their huge mouth agape like a whale shark. They are one of the largest freshwater fishes in North America, exceeding five feet in length and weighing more than one hundred pounds. Paddlefish are slow to mature, and females don't breed every year. Native to the Mississippi River Basin, they are known to make long migrations of as much as two thousand miles. Once common in twenty-two states, they have vanished from eleven and are endangered in several more. Overfishing, sedimentation in spawning areas, and development, especially dams, have caused declines.

In north Louisiana, paddlefish are called "spoonbill catfish" although they are not a type of catfish. The Cajun fishermen around Lacassine called them "spatule" or "belle dame," and because they have a cartilaginous skeleton instead of sharp bones they were relished as food for children. These local folks were the first to raise the alarm when the assault on paddlefish began, resulting in a closure of the commercial fishery in 1986. Today the commercial harvest of paddlefish and their eggs is still prohibited in Louisiana, but recreational anglers can take two fish per day with some restrictions. Paddlefish remain a unique component of our diverse aquatic fauna. Barring political complications, their future appears bright.

2
WORKINGS

Swamp Sleep

Swamps sleep naked and are slow to awaken. Long after green-up in the uplands, deep overflow swamps that sustain Louisiana bayous and rivers remain quiescent, prolonging winter dormancy until the threat of natural spring flooding has past. The palette of colors is subdued—the grays of barks and the browns of the forest floor. Only two exceptions are permitted in the spring canopy—clumps of dark green mistletoe in the treetops and mid-story sprays of dazzling, white mayhaw flowers. The fixed mistletoe is unchanging, never comes to earth. The mayhaw must go about its business of blooming early to accommodate an evolved relationship with shiny black bees no larger than a grain of rice. Pollination is a crucial matter not to be flirted with even if an occasional crop is drowned. While cherrybark oaks and sweet pecans on the adjacent ridges are fully feathered, overcup oaks and bitter pecans in the swamp are bare of new growth. Swollen buds are the only signs of the flush to come. In eons past, some of the swamp trees no doubt tried to get a head start on their neighbors by sending forth chlorophyll-laden leaves early in order to crank the engine of photosynthesis. Maybe it worked for a year or two. Maybe they grew taller than cohorts by capturing their sunlight for a while. But a swamp is a swamp for a reason, and the reason is cyclic pulses of life-sustaining water. To

Swamp Sleep

be a botanical player in the swamp a plant must adapt to the recurring cycles, and growing leaves too early eventually leads to a bad end. Leaves submerged under flood waters are lost and, though they often regrow when waters subside, the tree is stressed. Stressed trees are poor competitors and usually lose the game in time. They have no progeny. Better to be patient and sleep late.

Alligator Aberrations

When it comes to the subject of alligators, every detail of their natural history seems to be newsworthy. Two rare genetic manifestations of alligator skin have attracted thousands of visitors to zoos and other reptile exhibits in recent decades. The sensations began when eighteen white hatchlings were found on a nest near Houma, Louisiana, in 1987. They were not albinos but rather leucistic alligators. An animal with this condition has defective skin cells that are incapable of making any color pigments on all or part of its body (except for the eyes). In contrast, albinism is a genetic trait that inhibits the production of only the dark pigment melanin. Another difference between albino and leucistic animals is eye color. The eyes of albinos appear red because the internal blood vessels are not masked by a dark pigment, but those of leucistics are normal. Some of the captive Houma alligators are still alive and exceed ten feet in length.

Albino alligators have surfaced only slightly more often than leucistic ones. The acclaimed herpetologist Raymond Ditmars wrote in 1908, "There is an albinistic specimen living in the New York Zoological Park." An alligator farmer collecting wild eggs near Myrtle Grove, Louisiana, in 1991 found a clutch that hatched seven albinos. Over the next several years he acquired albinos from two nests in the same area each season. His conclusion was that there were at least two local females and one male alligator that carried the recessive gene for albinism. As recently as 2012, an albino alligator was exhibited at the National Aquarium in Washington, D.C. Curators there stated, "Fewer than 100 of these extraordinary species exist worldwide due to the many environmental challenges that they face." White alligators, whether albinos or leucistics, have no chance of extended survival in the wild. Their skin is sun-sensitive, and the stark coloration highlights them as easy targets for abundant predators. Those captured and exhibited for our amusement are the fortunate ones.

Green

Theologian Martin Luther once said, "For in the true nature of things, if we rightly consider, every green tree is far more glorious than if it were made of gold and silver." His spiritual adage was more literally accurate than he could know in the science of the early sixteenth century. Without Luther's green, oxygen-breathing organisms would not exist on planet earth.

In Louisiana, greens are most striking in spring as our landscape transforms from the somber hues of winter to the verdancy of new growth, and rare is the artist's palette that holds the countless

Fern

shades of green that emanate from a fresh forest. Marketers label color charts with the likes of "apple green," "lime green," and "jungle green" in shallow efforts to describe the natural world. More accurate would be "tendril green" of unfurling greenbrier vines, "hazy green" of an emerging white oak canopy with leaves the size of a squirrel's ear, and "braird green" as of the first peach-fuzz fronds of cinnamon fern.

The source of all plant greenness is a molecule called chlorophyll that is green because it reflects the various tints of green in the electromagnetic spectrum. Chlorophyll has the remarkable ability to capture sunlight, harnessing its energy to transform carbon dioxide and water into glucose for the plant's food, while releasing life-sustaining oxygen as a byproduct. The process, as every fifth grader should know, is known as photosynthesis. It is interesting that the chlorophyll molecular structure is remarkably similar to that of the human hemoglobin molecule and differs only by having a central

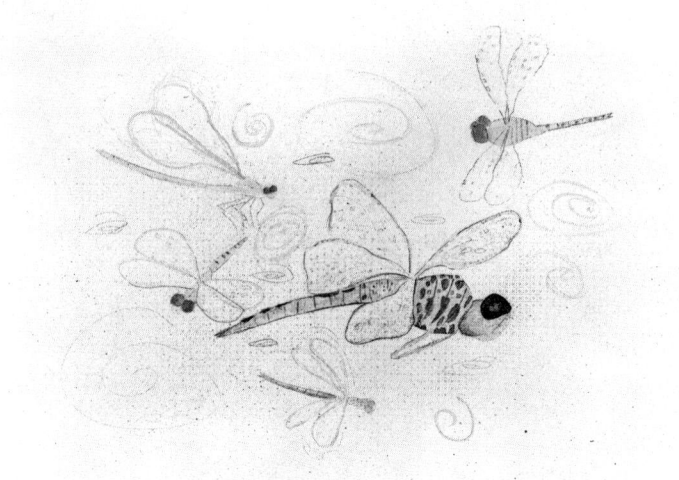

Dragonflies

atom of magnesium instead of iron. In this regard, we are not as far removed from plants as most would think. In every regard, we are utterly dependent on them and their greenness for every breath we take.

Phenology—A Heartwood Sampling

Phenology: periodic biological phenomena that are correlated with climatic or seasonal conditions.

19 September: Large flocks of common grackles, like rolling clouds of raucous ebony raiders, descend upon the willow oaks of D'Arbonne Swamp to pinch the meats of ripe acorns—a sign of changing seasons.

20 September: Two red-bellied woodpeckers hitched backwards down a large tree huckleberry to muscle chickadees out of the birdbath for a drink. Average precipitation here for the month of September is 3.5 inches. Ten days to go and we've received .05 inches in the Heartwood gauge.

22 September: The autumnal equinox for the Northern Hemisphere officially begins at 10:13 CST tonight. For the first time in nineteen years, a full moon will shine (at 4:17 a.m. on September 23) on the beginning of fall. Traditionally, the full moon nearest the autumnal equinox is called the harvest moon—and you can't get much closer than this one.

24 September: Before sunrise this morning a barred owl tipsy in the light of a barely over-ripe harvest moon reminds the local girls not to forget about him during this time of seasonal separation.

26 September: Almost an inch of rain fell last night before the first cool front of autumn, but there's nary a puddle in sight this morning. Activated root hairs sucked it all away into chambers of xylem, and transpiration pumped it upwards to the thirsty tips of the tallest trees. This silent process that sustains plant life is as vital to our own well-being as a heartbeat.

28 September: It's just a theory. For Louisiana, the relative humidity is low today. As he trolls the sky in erratic loops and sifts the wind for a molecule of decay, turkey vulture wishes for a bit more moisture in the atmosphere to fine-tune his receptors.

29 September: Common green darner (dragonfly) patrols his territory in the dry swamp with a vengeance. Normally, he guards pools in the shrub-swamp wetlands for his egg-laying mate. Is he expecting rain?

2 October: At dawn this morning the jalapeño pepper crop was harvested. After processing into pepper jelly and pepper sauce, the seeds and trimmings went into the compost pile. Possum in her nightly visit to this banquet table may have a difficult time maintaining her fixed grin tomorrow.

8 October: A coyote lopes across the pond levee this morning look-ing back over his shoulder at me with the insolence of a hor-mone-laden teenager. Also, in the form of a shedding summer coat he wears his pants too low.

10 October: The drought is persistent and a grim reaper for native trees at Heartwood. Already, dogwoods, wild azaleas, witch-hazels, some oaks and hickories have succumbed. Birds are more desperate than I knew, as even my front-porch banjo picking did not frighten blue jays and red-bellied woodpeckers from the oasis of nearby bird baths.

12 October: Strawberry bush, wahoo, American euonymus—not at this place. It's hearts a bustin' (with love) at Heartwood. This specimen doesn't appreciate her good luck. Just a few feet farther from the house she would be browsed to nubs by the resident does and their young of the year.

15 October: This morning a freshly migrated northern harrier (aka "marsh hawk" to old biologists) hunted voles in the yellow nut-grass fields of a nearby waterfowl sanctuary. The bird appeared to be an adult female. Those of her fair sex always arrive on the win-tering grounds before the adult males in their sleek gray plum-age. Perhaps it is because voles taste a bit like chocolate.

17 October: Narrow-leaved sunflowers exhibit heliotropism as their flowers follow the sun from east to west throughout the day—but only when they are young and pliant. Once petioles age and stiffen, they watch with fixed gazes for the first frost of autumn.

30 October: On a recent night and in rapid succession, three bolts of lightning came to earth within a hundred feet of this house. One killed the computer that administered my website. Backups aside, it was a mundane tragedy. I mourn though for two white oaks and a mockernut hickory that germinated when Ulysses S. Grant was president and whose sap of life was boiled away at the speed of light.

3 November: Almost three inches of slow rain broke the back of the drought at Heartwood. Organisms within kingdom fungi have been waiting for this event in order to go about their fructuous affairs of procreation. Overnight they appear as Halloween-orange sentinels on rotten logs and creamy discs of basidia scattered about the oak litter. A physiological ripening will occur in a few days, often ending in an explosive discharge of spores. Then it's back to the serious subterranean business of cycling essential nutrients and decomposing organic matter that I might exist.

5 November: Shorter days of autumn inhibit the production of green chlorophyll in black gum leaves, and unmasked red pigment that has been waiting patiently behind the scenes since spring now has a few bright days of glory.

9 November: The season of dragonflies is ending. This lingering yellow one that lands on the tip of my fishing pole is the same species found at elevations above eighteen thousand feet in the Himalayas, in Madagascar, Japan, and the Brazilian rainforest. Known as the Wandering Glider, it is the only dragonfly that inhabits remote Easter Island. Humans don't have a monopoly on adaptation.

24 November: All day yesterday, skeins of snow geese passed over this swamp, the largest movement of the season. They come from barren Arctic places with names like La Perouse Bay. Trees do not fit into their life cycle, so there is no reason to linger here. In a few hours they will hear their kin and then see them massed in the rice fields of southwest Louisiana where prairies once reigned. Like feathered, slow-motion tornados they will settle there for the winter, thus punctuating the mystery of migration.

29 November: On these dark nights they swim unnoticed down our bayous and rivers bound for a procreative rendezvous thousands of miles away in the Sargasso Sea. Only those American eels several years old and sexually mature feel the tug of the cosmos in every cell. For them it is now a one-way trip to their natal, spawning, and burial grounds. For us it is an enigma too profound to explain with science.

6 December: Frozen Fright—Bayou grackles bedizened in sunlight inspirit a sandbar during daily ablutions. This vulnerable occasion requires vigilance. Was that the shadow of the small lightning hawk?

20 December: Recurring periodically in this swamp, annual Christmas Bird Counts can be considered ritualistic phenology. A static circle with a diameter of ten miles is superimposed on the landscape. Armed with the best polished glass of Nikon and Zeiss, a couple dozen of the faithful search every niche to tally all birds in the imaginary ring. At day's end around a black pot of brumal chili the important questions are asked. Have we ever counted so many rusty blackbirds? Where are the robins this year?

Bark

In humans and other animals a covering of skin serves various functions, including protecting the body within. In trees and other woody plants, bark can be considered analogous to skin. Like skin, bark is comprised of several layers, some living and some nonliving. The outermost layer is called cork and does not consist of living cells. It is usually impermeable to water and gases. Moving inward, specialized layers of living cells perform critical functions, including the transport of nutrients. The nutrients are manufactured via photosynthesis in the leaves or needles and flow through sieve-like tubes throughout the rest of the plant.

Humans have been using bark products for thousands of years. The inner bark of some plants is edible. The spice we call cinnamon is finely ground bark of the cinnamon plant. Latex and resins are bark products used in chemicals. Tannin from oak bark was used to tan animal skins for centuries. Lifesaving medicines such as quinine and aspirin were made from bark. As a construction material, bark

Bark

is used as shingles and flooring. Native Americans made birch bark canoes, and today we grind it up to use as landscape mulch.

This discussion of bark would be incomplete without mentioning how we thoughtlessly abuse it even while cherishing the plant it protects. The invention of the gasoline-powered string trimmer has resulted in the unintentional and untimely deaths of countless landscape trees and shrubs. If a string trimmer has been used in your yard, I challenge you to look closely at the base of your woody plants. There is a very good chance they have been partially or completely girdled. Once the bark of a plant has been seriously damaged, the plant will never thrive to reach its potential and will often die. Besides destroying the nutrient transporting cells, bark wounds are prime entryways for pathogenic bacteria, viruses, and insects.

To view another bark-related travesty in our region, visit a public campground and consider the nearby trees that we love for their shade and aesthetic values. Most will likely be hacked, scored, burned, or carved with initials. Beech trees in particular are con-demned if they are so unfortunate as to germinate in a public area. Considering the many benefits that we have reaped from bark over the centuries, what does this unnecessary destruction say about us?

Drought

From our place on the edge of a Louisiana swamp I can smell the drought. The usual organic brew of odors is absent. Now it smells like northern New Mexico in early autumn—like a toddy of weath-ered adobe and rabbit-bush resin. It is late October, and we have had half an inch of rain since the fifth day of July. NOAA's Drought Se-verity Index considers this area of the Bayou State in the category of exceptional drought—the same classification as the tinder box that is southern California.

The impacts are compounding daily. There is mortality in my yard—a bed of Christmas ferns, a feeble red oak, and a sourwood I planted twenty years ago. Those dogwoods that have survived the anthracnose are on life support. The bayou down the hill from my house languishes currentless at pool stage. On my walk there this morning the heavy clay soil was cracked into puzzle pieces. At least a dozen cat squirrels raced across my path after drinking at the wa-ter's edge. Needles on the cypress trees are oranging prematurely. I watched a dishpan-sized snapping turtle root in the mud of a small, clear pond like a Guinea hog as his world evaporated by the moment.

This drought is not without precedent. One of the worst recorded in Louisiana occurred in 1896 when part of the state was rainless for almost seven months. Wells went dry. Truck crops failed, and it took ten acres to grow a bale of cotton. Farmers in the hill country drove their cattle to Lake Bistineau and then it dried up. The *New York Times* and the *Chicago Tribune* recorded the disaster.

As I write, cumulus clouds in promising shades of charcoal drift in from the east. But they are wayworn, lacking intent, and serve only to confine the humidity at ground level. They are nature's mockery of human vanity, like the high-water marks fifteen feet up the trunks of willow oaks on my morning walk.

Eyeshine

In my family there are stories about lean times during the Depression when rabbits were a welcomed source of protein in the household larder. Most were shot at night with the aid of a carbide lantern. Rabbits were detected by their eyeshine in the dim glow of the light. Boys, new to the venture, were reminded that because rabbits' eyes are on the side of their head, only one eye could be seen at a time. And if, when walking through the lonely swamp at night, a person were to detect a creature with two eyes shining, he should remember that such physiology is a trait of many predators that can see much better at night than a mere boy.

The cause of much hope and apprehension during these undertakings was a cluster of highly refractive crystals behind the retinas of the shining eyes. Known as *tapetum lucidum,* these organs make the pupils of some animals appear to glow when struck by an outside light source. Animals with the brightest eyeshine usually have more rods and fewer cones in their retinas, resulting in excellent night vision but also color-blindness. Not all animals have a tapetum or eyeshine. Humans don't. Those animals that do have eyeshine tend to be mostly nocturnal and include many mammals but also spiders, some fish, frogs, and alligators. The color of eyeshine also varies by species. Horses have blue eyeshine, fish have white eyeshine, and that of the possum and many rodents is red. The eyeshine of cats and canids, which include cougars and wolves, is yellow, a fact not lost on my hungry kinfolks when they spotted two glowing orbs in the heart of D'Arbonne Swamp.

Feathers

What do a chickadee at your Louisiana bird feeder, a tyrannosaurus that lived in northeastern China 175 million years ago, and a tragic

Feather

sixteenth-century play have in common? That the chickadee is covered with feathers is not surprising, but finding the richly detailed plumes on the fossil of a Jurassic dinosaur seems a bit incongruent. Feathers are made of a special group of proteins called keratins. During development, the proteins bond into twisted sheets that result in microscopic structures similar to but stronger than those found in the hair, claws, and horns of mammals. The job of feathers on modern birds is to provide insulation from cold temperatures in both air and water. Proto-feathers found on dinosaurs served the same function. Feathers allow birds the remarkable concept of sustained, controlled flight, a phenomenon shared only with bats and some insects. Feathers also play important behavioral roles in the lives of birds during courtship and defense of territory.

The significance of feathers for humans is cross-cultural and spans the globe. They have adorned the bodies of British queens, Aztec kings, and New York socialites. They are used in the religious ceremonies of Native Americans and in snakebite medicine by East Asians. Feathers are fashioned into fishing lures for anglers and regimental headdresses for generals. They have been stuffed into mattresses during times of peace and fletched on arrows for war. Feather quill pens yielded the U.S. Constitution, the novels of Jane Austen, and the complete works of Shakespeare. Paleontologists consider the discovery of feathered dinosaurs further evidence of kinship between those reptiles and the ancestors of birds—thus the connection between an oriental tyrannosaur, a Carolina chickadee, and incidentally *Romeo and Juliet*.

Fibonacci Numbers

Fibonacci numbers are integers of a mathematical sequence that begin with 0 and 1, and each subsequent number is the sum of the previous two. Thus the sequence starts as 0, 1, 1, 2, 3, 5, 8, 13, 21, 34, 55, 89, 144, and so forth. Named after a thirteenth-century Italian mathematician, Fibonacci sequences are popular in art and weaving design but also common in biological settings and even considered by some to be a basic "law of nature." In the realm of plants, two consecutive Fibonacci numbers are found in the arrangement of leaves on a stem, in the order of branching in trees, the spiral of an uncurling fern frond, and the arrangement of bracts within a pine cone. The number of flower petals for many plants is a Fibonacci number. Calla lilies have one petal, Louisiana irises have three, but-

Fibonacci Numbers

tercups have five, delphiniums have eight, ragwort has thirteen, and so forth.

In the animal kingdom, Fibonacci sequences appear in the biology of honeybees, in the geometry of some spider webs, and in the spiral design of seashells. The anatomy of such diverse creatures as porpoises, starfish, and humans has been shown to exhibit Fibonacci characteristics.

The phenomenon is usually attributed to evolutionary efficiency of design. Plants and animals don't know about this sequence; they just grow in the most efficient ways. Also, there are many exceptions to the Fibonacci sequence throughout the landscape of natural history. Primroses have four petals, and trout lilies have six. Instead of a "law of nature," Fibonacci events may just be a trend of the times—the geologic times.

Fish Migration

Even the most nature-deprived urban dwellers among us are aware of the basic concept of bird migration. They know that some types of birds fly north in the spring and return in the fall. However, few people, including most outdoor-oriented folks in this region who should know better, realize that migration is also a vital part of the life cycle of other kinds of local wildlife. Consider freshwater fish, for example. If we have an idea of fish migration, it is likely tied to iconic species like salmon of the Pacific Northwest. But in Louisiana too there are fish whose lives depend on their ability to migrate, often long distances.

American eels are a type of fish still found in many of our rivers and bayous. Their behavior is opposite that of salmon in that they breed and hatch in the Sargasso Sea area of the Atlantic Ocean and swim thousands of miles to live most of their adult lives in freshwater. Prehistoric-looking alligator gar move out of the deep water of big rivers to spawn in side channels and connected backwater areas. American paddlefish, often called spoonbills, also migrate to breed and must have large amounts of flowing water in order to reproduce. Another ancient species, the pallid sturgeon, is a big-river fish that once swam freely throughout the Mississippi River basin to feed and reproduce.

All of these fish are greatly reduced in numbers compared to historic populations. The pallid sturgeon is a federally protected endangered species, and others may not be far behind. Several factors are at play in the declines, but one is paramount—dams and other barriers. More than six million man-made barriers in America's waterways have contributed to the loss and decline of many species. Levees prevent gar from accessing breeding areas. Locks and dams reduce natural flowing currents needed by spawning paddlefish and sturgeon. Reservoirs block the natural migration of eels.

So why is the loss of these fish important? Some are not favored food species. Migratory fish are environmental indicators that gauge the health of our waters, waters that are critical for other wildlife and humans too. Alligator gar and pallid sturgeon are apex predators and thus key natural links in the life cycles of other species, some of which we are fond of eating. They are also a part of our cultural heritage and history.

The next time you hear the calls of migrating geese, remember that they are not the only creatures trying to traverse the anthropogenic maze of obstacles on the landscape.

Irruption

One of the joys of bird-watching is the ever-present chance of seeing something new—an unusual or perhaps rare species that suddenly appears unexpectedly. This is possible of course because many birds are great travelers, often flying thousands of miles in the mysterious phenomenon of migration. Especially in winter, bird-watchers across the country hope for the arrival of avian visitors that don't normally occur in their areas. When numbers of them do visit periodically, the phenomenon is termed an irruption. In North America the species most often associated with winter irruptions include pine and evening grosbeaks, crossbills, purple finches, pine siskins, and even snowy owls. Some of these birds usually spend their winters no farther south than southern Canada or the northern United States. In the Bayou State we experienced one such irruption in January 2013. Red-breasted nuthatches abandoned their normal haunts in northern coniferous forests and appeared throughout Deep South states in this winter. They didn't come for the sun and sand, and their arrival was not a portent of a harsh winter ahead. The cyclic occurrences are driven by a scarcity of food on the normal wintering grounds. Red-breasted nuthatches came to Louisiana because of a cone-crop failure in northern pines, spruces, firs, and larches. The small, short-tailed birds are dependent on seeds in the cones. Happy bird-watchers reported seeing them all over the state, some at bird feeders stocked with sunflower seeds. Nature's tendency is to flow in cycles, so when it happens again watch for these feathered Yankees while they're here as it may be years before they return.

Lunar Cycles

For as long as humans have looked up into the night sky, the moon has caused people to behave in strange ways. One of the most peculiar is the tendency for people to correlate phases of the moon with various human behaviors. Lunar cycles have been claimed to affect homicide rates, traffic accidents, suicides, the birth of babies, assaults, emergency-room visits, casino-payout rates, and psychiatric admissions. The only crazy thing here is that scientific studies have failed to show any reliable significant correlation of these events with lunar phases. The persistent belief in lunar myths seems to be tied to media effects, folklore and tradition, and general misconceptions.

The phases of the moon depend on its position in orbit around the earth, and the earth's position in orbit around the sun. Each phase has a different name depending on the amount of illuminated surface that can be seen from the earth. As the moon waxes or appears to grow it moves through the new moon, crescent moon, first-quarter moon, gibbous moon, and full moon phases before returning through the gibbous moon, third-quarter moon, crescent moon, and new moon phases. One complete cycle is exactly 29.53 days. Repeated studies consistently show that the average human menstrual cycle is also 29.5 days. Is that a coincidence?

Within the animal kingdom there *are* examples of behaviors tied to lunar cycles that have passed the test of scientific scrutiny. Corals spawn at full moon. Spotted owls call more frequently than would be expected in the third-quarter and new moon. Lobsters are more active in the dark of the moon. Galapagos fur seals vary their dive patterns according to the moon phase, and some sea turtles nest and lay eggs on a lunar cycle.

Humans have labeled plants and animals with names like moon vine, moonfish, moon weeds, moon rat, moonseed, moonwort, moon snail, and luna moth. My Uncle Charley plants potatoes and prunes shrubs only during a new moon. In fact, the connecting threads between these entities and lunar cycles may be more ephemeral than moonbeams.

Predators

The biologist's definition of "predator" is an animal that kills and eats other animals. In the natural world, predators are a very important part of ecosystems as they affect other species in the web of life both directly and indirectly. It is a misconception that predators always control populations of their prey species. In fact, it is often the other way around. Arctic foxes and snowy owls depend on small rodents called lemmings as their primary food source. Lemming populations naturally cycle through high and low periods. Populations of foxes and owls vary according to the availability of their prey. It is the lemmings that regulate the predators in this case.

Predators are found in all groups of animals—from red wasps to hognose snakes, to bluegills, to red-tailed hawks, to domestic cats. In many situations, predators themselves become victims of other predators higher up the trophic level. Predaceous insects are preyed upon by bullfrogs, which are eaten by black bass that are consumed by snowy egrets, which are killed by peregrine falcons. One group of predators, however, sits at the peak of the food chain and is not usually bothered by other predators except perhaps humans. These are termed "apex predators." In Louisiana examples of apex predators include black bears, bobcats, coyotes, bald eagles, ospreys, alligators, alligator snapping turtles, and of course humans.

From this list of apex predators and others such as sharks, wolves, grizzly bears, and cougars, it is obvious that the natural behavior of some is in direct conflict with the covetous conduct of many humans. They are considered competitors and often labeled vermin. These naive beliefs have led to massive eradication programs with unanticipated side effects. The role of predators in maintaining healthy ecosystems is significant and still being deciphered. We attempt to manipulate the grandiose complexities of life wearing blinders of blurred awareness and err in thinking that our well-being is somehow separate from that of the natural world.

Thyroid Complications

A recurring theme in *Bayou-Diversity* involves our connections and links to the natural world. Consider this hypothetical scenario. A young couple decides to celebrate their anniversary by dining out at a popular seafood restaurant on a warm spring evening. The special of the day is stuffed flounder, which they both choose to try along with a side order of fried frog legs as appetizers. When their dinner is served it is sprinkled with salt to embellish the rich natural flavors, and the meal is indeed memorable. Within the various elements of this setting there is a common biological thread that links them all. It involves your thyroid gland.

The thyroid is a butterfly-shaped gland wrapped around the windpipe just below the Adam's apple. It secretes several important hormones that regulate metabolic processes such as heart rate, blood pressure, body temperature, bone loss, and food movement through the gastrointestinal tract. The hormones are especially critical during infancy and childhood for proper growth and development, including that of the brain. The thyroid gland uses the element iodine to manufacture the vital hormones. To preclude iodine deficiencies and the ensuing thyroid-related diseases, iodine is added to most table salt.

As it turns out, humans aren't the only species with a thyroid gland as it is found in all animals that have backbones (vertebrates). However, the gland has different functions in different species. More than a hundred years ago a scientist fed ground-up mammal thyroids to tadpoles and observed that they immediately developed into adult frogs. Their external gills disappeared, legs and eyes grew rapidly, and the tail was resorbed. When the scientist removed the natural thyroid from the tadpoles, they never developed into frogs. So, the thyroid controls metamorphosis in frogs that results in fried frog legs on the platter. The stuffed flounder is entangled also. Flounders are a type of flatfish that also experiences radical developmental changes. When young, they are shaped as typical fish with eyes on opposite sides of the head. As they grow, one eye migrates to the other side, which becomes the top of the fish. This metamorphosis is also driven by hormones from the thyroid gland.

Table salt laced with iodine, stuffed flounder, fried frog legs, and your thyroid gland—it's complicated out there.

Firewood

This year a spring storm determined my source of domestic heat for the coming winter. Straight-line winds toppled a huge southern red oak on an upland hardwood slope a few hundred yards west of my house. I came across the fallen tree soon afterwards during my ramblings and mentally marked it for mid-summer firewood gathering. The trunk was straight and void of forked limbs for the first forty feet. In its prime the tree was 105 feet tall and twenty-eight inches in diameter—not a goliath but impressive nonetheless and indicative of having the good fortune to germinate on a moist, fertile site.

I often wish that such organisms could possess and share an anamnesis, the remembrance of past events in their lives. The life-

Firewood

span of this tree was on a human scale. It lived for eighty years, the same as my father. Indeed, Dad was a stripling and the tree was a sapling at the same time in the Great Depression. Their paths almost certainly crossed as both grew to maturity with roots in my great-grandfather's land on the edge of the D'Arbonne Swamp. The tree would know of the people's lives, of how the women gathered at the nearby spring on washday to fire up the black pots and scrub the hard work out of faded overalls with lye soap. It felt the thumping vibrations of great-grandfather's small, steam-powered cotton gin beginning each October with the first bite of frost on its highest leaves. After several years, autumn was quiet again as the shallow, red-clay soils were milked of nutrients needed by hungry cotton, and the gin boiler was hauled off in the scrap metal drives of World War II. Four generations of boys stung the tree and its neighbors with the prick of .22 rifles in the relentless pursuit of cat squirrels. An old fire scar on the butt of the tree may indicate someone's carelessness, or maybe just the common practice then of firing the woods to keep down the snakes and ticks. Whichever, the tree would likely know as it leaned away from the scar for many years.

The firewood is gathered now. Large blocks of the bole are piled and ready for splitting just off the front porch. When I walk past them every morning on the way to the garden, they reek of the sour, earthy smell of green oak wood. During the harsh winter nights to come, warmth generated by the remains of this organism will fuel *my* vespertine remembrances—recollections that include those in common with the tree.

Cosmopolitan Animals

In biology, references to an animal as "cosmopolitan" in range mean that it is found in most places around the world. Such a creature is considered a generalist in that it is usually adapted to survive in a wide variety of habitats. So what wild animals in the Bayou State also live in faraway places? In the bird world, peregrine falcons and barn owls fall into this category, along with cattle egrets and their amazing story of rapid-range expansion. Originally native to North Africa and parts of Asia, cattle egrets began expanding their range into southern Africa in the nineteenth century. By 1900 they had flown across the Atlantic Ocean and settled in South America. North America and Australia were invaded in the 1940s, and soon afterwards Europe was colonized by those birds still in northern Africa. Cattle egrets have a life history closely tied to large browsing animals that in their movements stir up insects for the birds. The dramatic spread in range was correlated with growth in livestock production around the world.

People have also facilitated the spread of several species of animals that would top a list of undesirables. Notably, the brown rat, house mouse, house dust mite, and German cockroach have followed us from place to place and are now almost ubiquitous. None of these species were in Louisiana before Euro-Americans arrived.

Like the barn owls and peregrine falcons, though, another species spread around the world without our help. The yellow dragonfly that lands on the tip of your fishing pole is the same species found at elevations above eighteen thousand feet in the Himalayas, in Madagascar, Japan, and the Brazilian rainforest. Known as the wandering glider, it is the only dragonfly that inhabits remote Easter Island. Humans don't have a monopoly on adaptation.

Names I

The term "ichthyologist" refers to a scientist who studies fish. Some ichthyologists spend their professional careers sorting out the classification of fish and labeling them with scientific and proper common names. Using a standardized system to assign names to plants and animals is essential to ensure that researchers around the world know exactly what species is being discussed without confusion. Nonacademics, on the other hand, are often acquainted with plants and animals in their surroundings and have always tagged them with their own descriptive colloquial names. Fish are no exception.

If a renowned German or Japanese ichthyologist with expertise in global fish taxonomy were to engage a local fisherman in a conversation concerning the aquatic species in our north Louisiana rivers and bayous, I'm afraid chaos would ensue. How could the expert know that the belittled scissor-bill and needle-nose refer to various species of gar, or that humpback blues and tabbies are esteemed types of catfish? For him, grinnel would not resonate with bowfin, and that spoonbills were actually paddlefish would be a mystery. As for the sunfish family, the local angler would mention white perch, goggle-eyes, snot-noses, and government bream as the man of science furled his brow. His only glimmer of comprehension would surface when a reference to top-water minners was close enough that it could be linked to the approved term "minnow." Imagine the confusion when he drove down the road to south Louisiana where the white perch transformed into a sac-o-lait and the grinnel became a choupique. That might be just enough to convert an ichthyologist into a mythologist.

Names II

There is a very good chance that everyone reading this book has witnessed a murder. The observation may have occurred in your own backyard. A murder you see is the collective term for a group of crows, as in "a murder of crows just descended on my corn patch." Many such archaic terms exist for animal congregations. Some seem to have a rational link to the species; others are outright bizarre. It begs the question: "Who comes up with these words?" In the bird world, examples include finches, a group of which is called a charm. Several owls are referred to as a parliament, a bunch of vultures is called a wake, and a few jays make up a scold. My favorite word for a bird group and one that makes perfect sense is that for cormorants. A group of these opportunistic scoundrels is called a gulp. I didn't make this up. In the realm of mammals, several ferrets are called a business, kangaroos and monkeys live in troops, three or more rhinoceroses make up a crash, and as we would expect, a group of hippopotamuses constitute a bloat. It's easy for us in Louisiana to relate to a nest of snakes or a shoal of bass, not so much for a shiver of sharks or a tower of giraffes.

I'm not sure that there is much value in learning these obscure terms unless you are a crossword-puzzle addict or writing the definitive historical novel about seventeenth-century England. Most should probably just remain archaic, not unlike a plague of locusts.

Bird Banding

Biologists and amateur naturalists have been banding birds in North America for more than a century. The process involves capturing wild birds and placing a metal ring on their legs. The rings or bands are inscribed with information that allows anyone who comes in contact with the banded bird to notify the bander. All such activities are now permitted and coordinated by the U.S. Fish and Wildlife Service bird-banding lab in Patuxent, Maryland.

Banding has revealed many remarkable aspects of the natural history of birds. For eons people noticed that some types of birds were present in their areas for only part of a year and mysteriously vanished at times only to return months later. Banding clarified the enigma of migration and revealed that birds often travel thousands of miles to exotic locations. As an example, banding showed that Arctic terns make an annual round trip of twenty-five thousand miles to and from their breeding grounds. We learned that some species fly south on one pathway and return north on another. Yellow warblers fly south across the Gulf of Mexico to Central America in the fall but return in spring via an overland route. Birds that nest in a particular location often winter together in specific areas. Snow geese that breed in Alaska winter in California. Those that nest on Hudson Bay often come to Louisiana.

Banding divulges the life span of birds and proves that some survive in the wild as long as twenty years, although most live for only a few years or less. Colored bands are used in some biological studies to determine basic life history such as local movements, size of territory, and behavior. Banding data are used to study diseases, especially when birds are vectors or reservoir hosts as in the case of West Nile virus. Sustainable hunting regulations for game birds are established with the aid of banding information.

Of course, bird banding is of no value unless some bands are recovered. This happens when banded birds are recaptured by banders, shot by hunters, killed by collisions with windows or cell phone towers, or otherwise salvaged. If you come across a bird band, be sure to report it. We can't know too much about feathered miracles.

Alligators and Cold Weather

The least-known aspect of alligator life history involves their behavior during the winter, especially in inland swamp habitat. In general, they retreat to dens during cold weather, but they do not hibernate. Instead, they brumate, a condition when the core temperature

Alligator in cold weather

and other physiological processes decrease, but not to the extent that occurs in true hibernation. Other kinds of reptiles, including some snakes and turtles, also brumate. Alligators must surface to breathe when brumating and apparently move in and out of this state as the weather changes. They bask on warm winter days, but an alligator out of water on a very cold day is usually the sign of a sick alligator. Their ability to slow down bodily functions allows them to survive cold weather only up to a point. Infrequently, extended periods of unusually cold weather, when water remains frozen for several consecutive days, occur in northern Louisiana and southern Arkansas. In the last thirty years I have observed alligator mortality within two weeks of almost all of these events. Usually the dead alligators were larger adults that floated to the surface. Larger individuals may have been more sensitive to cold or just more likely to be seen when they died. During a severe cold spell in the winter of

1983–84, thousands of alligators died in Louisiana, Texas, and Mississippi. The temperature dropped to 13 degrees in coastal marshes with an ice cover four inches thick for several days. Surveys showed that alligators of all sizes succumbed, and deaths continued for several weeks after the weather event. The extent of alligators' ability to regulate their body temperature limits where they can survive. As the climate warms in some areas, we might expect their range to expand, but there will always be setbacks.

Brer Swamp Rabbit and Brer Fox

This article is by my wife, Amy, who has written the children's book *Swamper: Letters from a Louisiana Swamp Rabbit* (Louisiana State University Press) that teaches the values of our wetlands from the perspective of a swamp rabbit. You will see the connection.

Pondering predator-prey relationships in the ecosystem is an interesting pastime. Animals in both categories are well adapted to survive in their environment. Survival often becomes a game of wits. Consider the swamp rabbit and the red fox. Each has a bag of tricks that will help it survive.

The swamp rabbit, a common denizen of bottomland hardwood forests in northeast Louisiana, has traits that have evolved to help it evade one of its main adversaries, the red fox. Swamp rabbits are well camouflaged. Their streaked black and brown fur blends into the surrounding vegetation. They have long, mobile ears that can detect the slightest noise. They have eyes on the sides of the head as do many other prey species, which gives them good lateral perception of movement. They also have a keen sense of smell and are wary, a trait vital to their survival. The first trick they use to prevent detection is to crouch and freeze. If discovered, they escape by running in a zig-zag manner that can be hard for a predator to follow.

The red fox also has a bag of tricks for survival. He has an acute sense of smell. Red foxes probably detect swamp rabbits by smell before they see them. Their ears are large and upright and can also sense the least noise. The red fox is a swift runner and has eyes on the front of its head that provide good depth perception. He runs in a straight line to chase prey.

So in this game of wits, who wins? Both species must win some of the time if they are to survive. Swamp rabbits have one trick that other kinds of rabbits don't have. They don't mind swimming. So, if the red fox is hot on his trail, he heads for the nearest body of water and dives in. He waits then with his nose poking out of the water. Red foxes aren't fond of water and may forfeit the hunt. If this happens, the swamp rabbit has outfoxed the fox. But if this doesn't work, then the rabbit may just plead "please, please don't throw me in the briar patch."

Wild Bird Longevity

While cleaning out wood duck boxes in anticipation of the upcoming nesting season, a biologist on Upper Ouachita National Wildlife Refuge recently found a roosting screech owl in one of the boxes. The small owls are often found in the boxes and are occasionally banded during the encounter. When this owl was captured, the biologist noticed that it was already banded. He duly noted the band number and released the bird. Upon checking records back at the office, he was surprised to find that the owl had been banded as an adult eight years earlier. This means the bird is now at least ten years old and brings up the question of how long birds live in the wild.

First it should be noted that records of captive birds are not relevant because of the natural hazards faced in the wild. It is hard to determine longevity of wild birds, and most reliable records are a result of recapturing banded birds as with the screech owl mentioned earlier. Within the lifespan of birds, the highest mortality occurs in young birds soon after they fledge. From that point forward, survivors face a never-ending threat of dangers, and very few die of old age. As an example, studies show that the average adult life span of Louisiana songbirds is about ten months. In general, larger birds live longer than smaller birds. Great blue herons live longer than killdeer, and bald eagles live longer than house wrens. The following list indicates longevity records for several species of Louisiana birds.

Species	Yr.-Mo.	Species	Yr.-Mo.
Canada Goose	23-06	American Robin	13-11
Mallard	23-05	House Sparrow	13-04
Great Blue Heron	23-03	Brown Thrasher	12-10
American Coot	22-04	Wild Turkey	12-06
Osprey	21-11	American Kestrel	11-07
Bald Eagle	21-11	Song Sparrow	11-04
Red-tailed Hawk	21-06	Black and White Warbler	11-03
Brown Pelican	19-08	Tree Swallow	11-00
Mourning Dove	19-03	Acadian Flycatcher	10-11
Sandhill Crane	18-06	Killdeer	10-11
Great Horned Owl	17-04	Dark-eyed Junco	10-09
Blue Jay	16-04	Scarlet Tanager	10-01
Hairy Woodpecker	15-10	Ruby-throated Hummingbird	9-00
Northern Cardinal	15-09	House Wren	7-01
American Crow	14-07	Golden-crowned Kinglet	5-04

Source: web.stanford.edu/group/stanfordbirds/text/essays/How_Long.html
(accessed February 6, 2017).

It's important to remember that the average life span for each species is much lower than these maximum records.

Another interesting aspect of the owl encounter at the Upper Ouachita refuge surfaced in the banding data. It turns out that the screech owl was recaptured in the very same wood duck box in which it was banded eight years earlier. He seems to have found a great place to live a long, healthy life.

Dendritic Patterns

Persistent patterns of nature permeate our bodies and our environment, and for the most part go unrecognized by all but the very observant. One ubiquitous design is the dendritic pattern. "Dendritic" refs to a shape that resembles a branched tree. It is a pattern that is associated with growth or movement. Consider how the main trunk of an oak forks into large limbs that fork again and again into smaller branches and twigs. The pattern continues in the veins of leaves and the vast complex of unseen roots. Our bodies and those of all animals are living assemblages of dendritic patterns. Ever-branching arteries, veins, and capillaries transfer blood through a circulatory system arrayed in dendritic form. Likewise, air passages in the lungs replicate the pattern, and the term "dendrite" describes the branched projections of neurons that comprise the nervous system.

Dendritic patterns are not limited to the biotic world as they frequent inorganic features as well. The best example involves water moving across a landscape. Imagine a satellite view of the Mississippi River basin. Large tributaries such as the Ohio, Missouri, Arkansas, and Red rivers join the big river at an acute angle. These in turn are fed by smaller rivers and bayous, which themselves receive the waters of lesser creeks and streams. An aerial photograph of the system clearly reveals a dendritic pattern. Snowflakes form in dendritic shapes, as does the frost on your windshield and crystals that grow on sedimentary rocks. One of the most dramatic examples of dendritic patterns is exposed in bolts of lightning.

Once the frequency of the pattern throughout the natural world has been recognized, the obvious question is "Why is this so?" in light of the many other potential patterns that could exist. Mathematicians explain the phenomena with fractals, special sets of numbers that define similarities at various scales. Fractals can reduce complex designs to a basic simplicity. The results are systems that function with the greatest efficiency. Dendritic patterns in plants allow them to maximize sunlight and transfer nutrients up and down most effectively. Likewise, the dendritic patterns in our lungs and vessels move air and blood most efficiently for our well-being, and water moves across topography, taking the path of least resistance in dendritic patterns. Humans, lightning, and oak trees—we are not so different.

Misused Biology Terms

If there was such a thing as word police to enforce the correct use of biological terms, jails would be full of repeat offenders. None of the violations rank as felonies, but misdemeanors are rampant. Here are a few examples. The terms "venomous" and "poisonous" are often misused. Rattlesnakes, cottonmouths, and copperheads are not poisonous; they are venomous. Venomous organisms deliver venom into other organisms, using a specialized apparatus such as fangs or a stinger. Other venomous animals include some bees, wasps, ants, spiders, and scorpions. On the other hand, poisonous organisms do not deliver their toxins directly. They usually have to be eaten or touched for the repulsive process to work. Monarch butterflies are poisonous to predators because they consume and concentrate toxic chemicals in milkweed when they are in the larval stage. The defensive chemicals can cause vomiting, blistering, or serious cardiac issues in predators that try to eat them. Poison ivy is correctly labeled a poisonous plant because we pay the piper with a contact dermatitis if we touch it.

If you want to make a biologist cringe, refer to the boney appendages that grow from the skull of male deer as "horns." They are not, but rather are correctly termed "antlers." True horns consist of a core of dermal bone covered by an epidermal sheath. The sheath is the actual horn, and they are not usually shed. Cows have horns. Antlers are branched structures of bone characteristic of the deer family and are shed annually.

Another example involves the term "extinction." People often relate that a certain animal is extinct in a specific geographic area, such as declaring that grizzly bears are extinct in New Mexico. By definition this is incorrect because a life form becomes extinct when the last remaining individual of that species on the planet dies. In the grizzly bear example, it would be more correct to say that grizzly bears are extirpated in New Mexico.

I must admit that there are some blemishes on my word-police record, but to date I have been pardoned and my goal is to avoid recidivism.

Those Sensual Plants

Does your magnolia tree in the front yard have feelings? How about the Better Boy tomato plants that you pinched the suckers from yesterday? In order to evoke feelings, it is necessary to capture some sort of stimuli from our surroundings, and for that to happen we need sensory receptors. In humans that means eyes, ears, a nose, taste buds, and touch-sensitive cells. But what about plants? Can they hear, see, smell, taste, and recognize touch? Twenty-five years ago these ideas were considered on the fringe of credibility; today plant science is revealing the concepts to be routine mechanisms of life in the botanical world. Since plants receive environmental cues

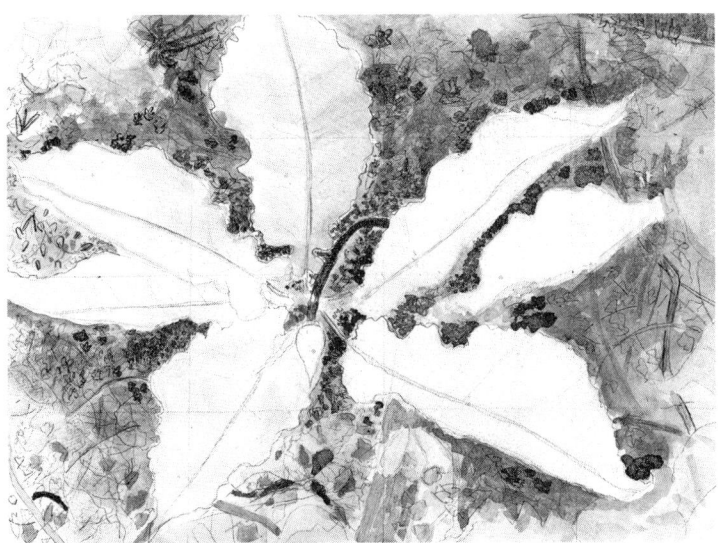

Big Leaf Magnolia

using organs different from ours, the definition of terms like "seeing," "smelling," or "tasting" has been broadened considerably.

Seeing involves the ability to perceive light. In humans that occurs in four types of photoreceptors in the back of our eyes. Plants have as many as eleven kinds of photoreceptors scattered throughout their leaves and stems. Consider the many types of sunflowers whose "faces" follow the sun across the sky on a summer day, or the canopy of a young oak tree that grows toward light and away from the shade of its neighbors.

The sense of touch in plants is expressed in many ways. The tendrils of vines such as English peas and poison ivy climb, cling to, and twirl around supports, using touch. Carnivorous plants like Venus flytrap snap shut at the touch of insects, and they are able to differentiate between an insect and a raindrop. As children we were amazed to see the tiny leaves of mimosa plants suddenly close up when stroked.

Plants can smell in that they can detect the volatile chemicals in odors. Tomatoes wounded by insects are known to emit odors that are alarm signals for their neighbors, who can then produce chemicals that defend against the attackers. If you put a ripe and an unripe apple in the same bag, the unripe one will ripen faster because it detects a ripening pheromone released by the other. In the wild it is often advantageous for a tree's fruit to ripen simultaneously in order to better attract animals that will disperse its seeds. As in humans, taste in plants is closely tied to smell but less understood.

Contrary to popular accounts, plants don't likely perceive music. There is no hard evidence that my mustard greens prefer Willie Nelson over Johann Bach. However, the chewing sounds of an insect are another matter. In one experiment, when researchers played recordings of munching caterpillars, nearby plants flooded their leaves with potent chemicals to deter the perceived attackers. The exact parts of plants that respond to sound have not been discovered, but a special protein found in all plants cells is suspect.

Plants have other senses as well. Some can perceive magnetic fields and color. By detecting gravity they can sense up from down; thus shoots grow up and roots grow down instead of in random directions. Recent DNA analyses reveal that plants and animals are more similar than most people might think. Why then wouldn't plants sense the world in similar ways?

3
TIMES PAST

Louisiana Catahoula Leopard Dog

With haunting "glass" eyes and varied coats of mottled hues, Louisiana Catahoula Leopard Dogs blend into the aura of Louisiana's mysterious swamplands—an important landscape in the breed's origin. They wander through Louisiana history in legends that involve Jim Bowie, President Teddy Roosevelt, and Governor Earl K. Long. Officially designated the Louisiana state dog in 1979, the breed is also called Catahoula Leopard Dog, Catahoula Cur, Catahoula Leopard Hound, Catahoula Hog Dog, and simply Catahoula. Regardless of the label, the breed is one of only a few that originated in the United States.

Factual claims of the breed's origin lack credible historic documentation. A common theory is that Catahoulas derive from a cross of Hernando de Soto's war dogs, during his destructive foray across the Southeast in the 1540s, and Native American dogs. Another hypothesis suggests that the breed developed much later, in the 1800s, when French settlers bred their Beaucerons with the Indian dogs. The name "Catahoula" seems to have been applied to the dogs by the early nineteenth century and indicates that the modern development of the breed has strong connections to the Catahoula Lake region of central Louisiana.

The general appearance is a dog of medium to large size, well-muscled and athletic. The coat is short to medium in length and diverse in coloration. A merle gene manifests in the "leopard" coats of Catahoulas and in the icy, colorless eyes of some individuals. Catahoulas are considered herding dogs and are also used to hunt a variety of small and large game from squirrels to bears. The herding trait was especially desirable to stockmen running semi-wild herds of cattle and hogs in the vast, unfenced Louisiana swamps. Owners and breeders report that Catahoulas make excellent pets but always require obedience training because of their assertive nature.

Anecdotal reports suggest that Jim Bowie favored the breed and even slept with a Catahoula at his feet. A pair of Catahoulas are said to have been members of the pack of dogs that pursued bears during President Teddy Roosevelt's 1907 hunt in northeast Louisiana. Governor Earl K. Long is known to have collected Catahoulas and to have used them to hunt feral hogs. In spite of the breed's time-shrouded pedigree and vague relationships, Catahoula Leopard Dogs are now firmly ensconced in the lore and law of the Bayou State. Act 239 of the legislature's 1979 regular session reads: "There shall be an official state dog. The official state dog shall be the Louisiana Catahoula Leopard dog, as registered by the National Association of Louisiana Catahoulas." Governor Edwin Edwards signed the legislation on July 9, 1979.

Carolina Parakeets

They were thought of as noisy mobs of rogues hell-bent on destruction. They swarmed the grain fields and orchards of European settlers, consuming the fruits of hard labor. If they possessed redeeming qualities it was only after they were dead and skinned, either for decoration on women's hats or fried in lard for the table. Linnaeus named them "Carolina parakeets" in 1758, and within that group there was a subspecies with slightly different-colored plumage called the "Louisiana parakeet." Carolina parakeets lived farther north than any other member of the parrot family. From southern New England to as far west as eastern Colorado and south to the Gulf Coast this small parrot inhabited hardwood forests, especially preferring the primeval bottomlands in the Mississippi and Missouri river basins. Within the modern boundaries of Louisiana this included most of the state.

Carolina parakeets were about thirteen inches long, with wingspans of just less than two feet. Mostly green with a yellow and orange head and splashes of yellow on the shoulders and down the wings, they were clown-like in appearance and behavior, living in noisy flocks of up to three hundred. At night they roosted in cavities of large, hollow trees of the swamps such as cypress and sycamore. During the breeding season they nested in similar holes. Wild fruits and nuts were their natural foods, including those of beech, maple, oak, pine, cypress, and grapes. Many historical accounts mention Carolina parakeets' affinity for cockleburs, a dreaded weed in southern fields. However, even their valuable service in controlling cockleburs did not mitigate the wrath of farmers who viewed their depredations as warranting death sentences. John James Audubon wrote, "The gun is kept at work; eight or ten, or even twenty, are killed at every discharge. . . . I have seen several hundreds destroyed in this manner in the course of a few hours." Audubon himself shot a Carolina parakeet near St. Francisville about 1821 to use as a model in his painting of that species.

Relentless killing of Carolina parakeets likely played an important role in their demise, along with deforestation of critical habitat in the nineteenth century. The last credible sighting of the birds in Louisiana occurred in West Baton Rouge Parish in 1880. The species likely persisted in Florida until about 1910, and the last known individual perished in captivity at the Cincinnati Zoo in 1918. Because they did not fit our mold as a desirable life form with which to share this planet at that time, we freed ourselves of the rowdy scoundrels for all times.

Wildlife Management—Hard Lessons

At the turn of the twentieth century the science of wildlife management was in its infancy. Reeling from the catastrophic human-induced losses of America's iconic fish and wildlife resources such as the vast bison herds and billions of passenger pigeons to market hunters, and countless plumed wading birds for the sake of vanity, a growing contingent of citizens began demanding a counteractive response to the wholesale pillage of nature. Thus was born the profession of wildlife management.

A common objective of many such programs was the restoration of depleted fish and wildlife species. Not surprisingly and because the field of ecology (the interrelationships of organisms with their environment) was yet to hatch, some well-meaning programs involved efforts to establish species in areas where they had never lived in the first place. A good Louisiana example involves elk.

Historically, elk were very rare, transient visitors in the state. Only one record of substance occurs in the literature. However, this fact was of no concern to early biologists and eager citizens who, knowing the popularity of elk as a western game animal, sought to add this species to the list of Louisiana's resident fauna. This sentiment resulted in the introduction in 1916 of wild elk from none other than the famous herds of the Yellowstone region. On January 24, 1916, forty elk captured on a ranch near Gardner, Montana, arrived on a train car at Urania in LaSalle Parish. They were promptly released on two thousand fenced acres of the state forest preserve. Ten of the animals died soon after release "from broken ribs puncturing the walls of the lungs" during the long, jolting cross-country transit. Later, one escaped from the enclosure and was struck by a locomotive. Another was shot by a local citizen, who was quickly prosecuted for the poaching. By 1920 only twenty-four animals remained, and soon after all were gone. They could not adapt to such a drastically foreign environment.

Although the experiment was a failure, as were many others across the country in this period (rainbow trout were unsuccessfully introduced in Tangipahoa Parish streams in 1917), some non-native introductions were remarkable successes, such as planting ring-necked pheasants in America's Midwest. Since then, the science of wildlife management has progressed and spawned the term "adaptive management," which means proactively evaluating and learning from our hard lessons of the past.

Turpentine

Until the middle of the twentieth century, few people in the South escaped an occasional medicinal dosing of a chemical derived by intentionally injuring native pine trees. The chemical was turpentine, and its uses were legion. Turpentine was derived from the resinous gums of pines, most often longleaf pine, also known as pitch pine, wherever it occurred. This resin contains the volatile hydrocarbon terpene. Turpentining involved chopping V-shaped notches into living trees and collecting resin that flowed from the wounds into boxes below. The cuts are called "catfaces" for their resemblance to a cat's whiskers, and the scars can still be seen today on some old trees in Kisatchie National Forest. Using copper stills not unlike those used in moonshine production, the resin was distilled through evaporation to yield turpentine and rosin.

Even though some turpentine was manufactured by individuals for home use, most involved large commercial enterprises. One of the largest operations in Louisiana was located near DeRidder and was comprised of three distilleries that averaged a weekly production of 150 barrels of turpentine, employing more than 250 workers. In this effort collection boxes were placed on 505,000 trees.

The uses of turpentine were almost limitless in number and creativity. Medicinally, it was used as a stimulant, diuretic, antiseptic, and laxative, often mixed with coal oil or kerosene. At various times turpentine was used in the treatment of worm infestations, epilepsy, tetanus, diabetes, yellow fever, typhus, tuberculosis, rheumatism, toothache, dysentery, and puncture wounds. Civil War surgeons during the heat of naval battles injected hot turpentine directly into wounds and coated stumps resulting from amputations with it. The treatment of many horse and mule ailments also involved turpentine.

Aside from medicine, turpentine was used in making soap,

candles, lighting oil, furniture polish, and rodent repellents. Today turpentine is still used as a solvent to thin oil-based paints but has largely been replaced by cheaper substitutes derived from crude oil.

Bayou Boats

For as long as humans have dwelled on our bayou-laced landscape, boats have drifted along the placid waters. Local Native Americans built watercraft for four hundred generations before European immigrants arrived to mimic their designs. For efficient travel and trade in a wilderness world of wetlands there were no other options. The earliest boats were dugout canoes or pirogues. Hewn from logs of virgin cypress or water tupelo, some were large enough to carry a dozen passengers or a thousand pounds of freight. Construction was labor-intensive and required skilled craftsmen, making the boats valuable assets. When settlers introduced pit-sawed lumber manufactured in crude sawmills, plank boats still in the shape of pirogues replaced the technically complex dugouts. The availability of lumber also allowed design diversity, and new types of boats were soon rounding the bayou bends. First were square-ended, flat-bottomed paddle boats called bateaus, "Joe boats" or "John boats." Skiffs had pointed bows for rougher water. Boats were often unique to a specific Louisiana bayou or river according to environmental conditions and their intended use. Even paddles and oars made of hickory or ash could often be traced to a particular community by their design. Early motor-propelled boats became common by about 1910. Sometimes called "gas boats," they were pushed by inboard gasoline motors and a long, propeller-tipped drive shaft. From this point forward the remoteness of many bayous was forever diminished. By the time of the great flood of 1927 the use of outboard motors such as those manufactured by Ole Evinrude was widespread, being portable and more efficient. Today, having succumbed to those soulless vessels of molded fiberglass and welded aluminum, wooden boats on Louisiana's inland waters are as scarce as the craftsmen who once transformed cypress logs into vessels of grace.

Alligator Rights

In Louisiana, matters concerning wildlife often end up in the courts. Usually they are criminal matters that pertain to the illegal taking of wildlife or related violations. Unique cases do crop up infrequently to embellish the oft boring legal records. One such example is reported to have occurred in 1850s antebellum New Orleans and documented in a *Harper's* magazine of that era. A summary of the case follows:

> A Miss Nel Gary went before one of the Recorders of New Orleans and made oath that one Ernest Dalfin, a neighbor of hers kept in his yard an alligator of immense size and ferocity; and that as she was frequently obliged to go through the yard, she considered herself in great bodily fear and danger; wherefore she prayed that her neighbor remove the alligator to some other quarters. On this charge, Dalfin was arrested. When required to plead, he stated that he kept the alligator to guard his premises from intrusion and that his guardian was, except when imposed on, as quietly disposed a reptile as ever lived. As for the prosecutor, he contended that she was brazenly inclined, and kept constantly exciting the alligator's ire by tickling him under his short ribs with ten foot poles and casting brickbats at his countenance and on one occasion even went so far as to singe his back with a hot smoothing-iron since which time his alligatorship swings his tail at her whenever he sees her. On this showing, Ernest was discharged; but Ellen was bound over to keep the peace toward "the pet" and its excellent owner.

It is unlikely that this case set precedent that was later tested in higher courts, but at least the monotony of cases involving routine mayhem in the tempestuous Big Easy was broken for a day.

Bayou Memories I

I have early memories of a bayou that became a prominent geographical feature in my life. Sometimes I even think I know quite a bit about the stream and its attendant swamp. This bayou was allegedly named for someone in a clan of Frenchmen who had ties

Red-eared Slider

to the region. Some historians point to Jean Baptiste Darban or d'Arbonne, the son of Jean-Baptiste d'Arbonne of Natchitoches. Others think it was the ancestor of these men, Gaspard Derbanne, a Canadian hunter who traveled with Louis Jucherneau St. Denis to the Red River country in 1714. In any event, it is now labeled on maps as "Bayou D'Arbonne," not to be confused with a lesser stream in St. Landry Parish called "Bayou Darbonne" but lacking the apostrophe in its correct spelling. With more than four hundred named bayous in Louisiana, an apostrophe is relevant.

The French and other Euro-Americans were latecomers on the landscape. If the bayou were uncoiled and stretched straight to depict a time line, the original people would be noted first in the headwaters of the upper reaches, drifting downstream for thousands of years until contact with Caucasians somewhere near White's Ferry, just a couple of miles above the mouth of the bayou. There, waters of the two cultures would roil with twisting currents for several hundred yards in time until the New World people evaporated as a race two centuries before flowing into the present.

Another approach to consider human time-in-place comparisons uses generational counts. Native people lived along Bayou

D'Arbonne for 400 generations. After contact, they were contemporary with people of European and African ancestry for four generations before disappearing. As latecomers, we have now been present about nine generations.

In the field of psychology, the term "genetic memory" is defined as that memory we are born with. It originates in the idea that over time common experiences of humans as a species become a part of our genetic code. If so, it can be said that my memories, which relate to knowledge concerning Bayou D'Arbonne, are about 390 generations short of that possessed by the last Native Americans to paddle this sinuous stream.

Civil War Witness Trees

The sesquicentennial of the American Civil War recently passed. Just over 150 years ago a conflict occurred that wrenched our country apart and killed more Americans than all combined wars since. Between 2011 and 2015 events around our reunited nation commemorated those awful times in hopes of learning lessons for today from behavior of the past.

Of course, no one is alive today to recount the history of the Civil War era. We are dependent on millions of records, books, letters, diaries, and other documents to narrate the stories, large and small. No people still live who beheld the drama, but there are still surviving organisms that trembled at the impact of mortars, lost limbs to cannon fire, and suffer musket balls deep within their living tissues. They are witness trees, trees that stood silent and witnessed the carnage of battles and persist in life today. Like the last Civil War veterans who faded into time in the twentieth century, the remaining witness trees are succumbing to age, disease, and accident. If the survivors could talk, their stories would enthrall.

A giant sycamore on the bank of Antietam Creek is a living witness to America's bloodiest day when on September 17, 1862, Confederate and Union armies fought a battle that yielded more than twenty-three thousand casualties. On the first day of the tide-changing battle of Gettysburg, Union General Daniel Sickles established his headquarters under a surviving swamp white oak. A cannon ball crushed the general's leg as he sat horseback in the shade, surveying the action. A silver maple at Shiloh beheld similar events, and a red oak at Bull Run stood through two major battles. President Lincoln delivered the Gettysburg Address near a huge honey locust that lives yet, and a pin oak at Appomattox Courthouse witnessed the formal end of the national tragedy. In Louisiana there is little doubt that some giant cypresses and live oaks that witnessed Civil War action still survive along quiet bayous while holding their secrets close.

In these persistent examples of natural history, American history is still with us.

Bayou Earthquake

Dr. R. F. McGuire was a prominent Ouachita Parish physician and planter in the first half of the nineteenth century. He was also a diarist and kept a journal from 1818 until 1852. An educated man, he dutifully recorded the weather and other natural phenomena along with business and politics of the era. His entry of April 7, 1842, hints intriguingly at links to a human tragedy that occurred fifteen hundred miles away on that same day.

At about 5:00 p.m. local time, several thousand feet below the island of Hispaniola in the Caribbean, where two giant plates of the earth's crust meet under grinding pressures, one of the plates suddenly lurched by twelve feet. The result was a devastating earthquake with an estimated magnitude of 8.1 and a dreadful tsunami. The earthquake was felt over a wide area, including Jamaica, Cuba, and Puerto Rico, but the northern coast of Haiti and what is now the Dominican Republic received the brunt of the natural disaster. Following the earthquake at Port-de-Paix, Haiti, the sea withdrew two hundred feet from shore and returned to drown the city in fifteen feet of water. Five thousand people perished there as the estimated human mortality throughout the region approached ten thousand. History remembers the event as the 1842 Cap-Haïtien earthquake.

Communications of the day precluded any chance of Dr. Mc-Guire learning about the disaster for weeks. However, in referring to that date he writes, "it is reported the waters in the [Bayous] Darbonne & [Choudrant?] were instantly [raised] about a foot with a gurgling noise & receded again without any storm." He goes on to speculate that the bizarre event may be tied to the presence of a comet. We now know this to be a false nexus, but other than the distant earthquake, what could explain the strange happenings on Bayou D'Arbonne?

Troyville Site

In Europe the time period between A.D. 400 and 700 was considered part of the Dark Ages. Foreign invaders ravaged the region, disease epidemics were rampant, and the population declined substantially. In contrast, Native Americans in the Lower Mississippi Valley were experiencing positive cultural and economic change and population growth at this time. A long history of constructing monumental earthworks for religious and political purposes continued. Deep in the midst of a vast bottomland hardwood swamp, one massive mound complex rose in present-day Catahoula Parish at the confluence of the Black and Little rivers. It consisted of at least nine mounds bounded by a D-shaped earthen embankment. At eighty feet tall, the principal structure was the second tallest mound in eastern North America. It became known as the Troyville site. When President Thomas Jefferson read of its discovery, he considered it so important that he briefed Congress on the matter. Today as an archaeological treasure nothing remains visible to the untrained eye. Its destruction began with the invasion of the ancestors of those who survived Europe's Dark Ages.

Serving as refuge for humans and livestock during years of high water, the larger mounds suffered from severe erosion. Confederate troops dug rifle pits in the tallest mound during the Civil War. The death knell for the remarkable site sounded in Baton Rouge when Governor Huey Long announced his ambitious statewide bridge-building program. The approach to the new bridge in Jonesville was in exact alignment with the large mound. A steam shovel leveled it in a matter of days.

No better example of wanton destruction of a world-class archaeological site exists in America. The owner of the largest mound reportedly sold its dirt to the bridge contractor for one hundred dollars. He likely had little choice. What was lost? Gone is the opportunity to learn more about the builders, who they were, their life ways, why they chose this site, and why they left. Today, an intact site of such grand scale would likely be a major tourist destination in an economically depressed region. The history of the Troyville site after Euro-American contact represents a sad case of how not to preserve and benefit from our cultural treasures.

[A version of this story first appeared in the *KnowLouisiana Encyclopedia,* 2014.]

The Dunbar-Hunter Expedition

The Lewis and Clark expedition to the Pacific Northwest is considered a hallmark in the exploration of the American continent. Conceived by President Thomas Jefferson to explore and document the geography, natural resources, and Native American cultures in the vast, unknown regions recently acquired in the Louisiana Purchase, this early nineteenth-century probe was not a solo act. Concurrently and with similar objectives, the president also planned a southern version of the now famous Corps of Discovery. Led by two prominent Scottish immigrants, William Dunbar and Dr. George Hunter, the venture was originally intended to trace the Arkansas and Red rivers to their sources. Had it succeeded as planned, it would have rivaled the Lewis and Clark expedition in scope. However, aggressive Osage Indians along these rivers altered the president's scheme, and he sent Dunbar and Hunter to explore the Ouachita River instead.

The party of nineteen departed St. Catherine's Landing below Natchez on October 16, 1804. To access the Ouachita River, the voyagers had to travel down the Mississippi River to the mouth of the Red River and proceed upstream to its confluence with the Black River and then the Ouachita. Along the way Dunbar and Hunter conducted their daily routine of recording their position along with a description of the region, including the biodiversity. Struggling upstream through frequent shoals that required loading and unloading of the boats by the recalcitrant crew, they poled into the frontier village of Fort Miro (now known as Monroe) on November 6. The party continued upstream and arrived in the vicinity of Hot Springs, Arkansas, on December 6. For a month the expedition camped at the remarkable cluster of thermal springs while Dunbar and Hunter conducted their scientific inquiries. They arrived back in Natchez in late January 1805.

The accomplishments of the Dunbar-Hunter expedition, though on a much smaller scale than Lewis and Clark's epic journey, were nonetheless noteworthy. They gathered geographic data on a wilderness region that was later compiled into an accurate map. They were the first to scientifically document the marvels of the hot springs. Their observations provide valuable insight into the flora and fauna of the Ouachita River Valley while still in a pristine setting. They verified for a young United States that the area was already well known and traveled by Euro-Americans, mainly commercial hunters and traders, thus offering the possibility of flourishing settlements in the near future.

Frisby Plantation

Standing on the bank of the Tensas River more than thirty years ago, deep within a swamp bearing the same Native American name, I could hear the hollow peals of a plantation bell—at least in my imagination. With no human habitation for miles on this subtropical summer day there wasn't much chance of it happening in real time, but a century and a half earlier the scenario was likely. Thank goodness it doesn't exist now for the real bell beckoned dreams of grandeur for one man and misery for many.

Just prior to the Civil War, Norman Frisby bought twenty thousand acres here in the heart of the remote Tensas Swamp with plans to establish a giant cotton plantation. As with all early Euro-American settlement in Louisiana, it was critical to locate on a navigable bayou or river to facilitate transportation in what for all practical purposes was a roadless world. The Tensas River met this requirement but only during the seasonal high waters when steamboats could navigate the winding stream. Frisby's big house was to be one hundred feet square, framed with virgin cypress, and rise three stories above the swamp. Slaves erected it on thirty-two columns made of bricks fired from riverbank clay. Outbuildings and a steam-powered cotton gin were built nearby. A large plantation bell, allegedly but improbably made of silver, was mounted on cypress timbers to mark time and control the lives of people in bondage.

Frisbee's grandiose plans collapsed in ruin with the start of the Civil War. The mansion was never completed, the slaves scattered to the winds, and inexorably the swamp began to reclaim the plantation. Frisbee's fate is clouded in time; one story has him killed by his brother-in-law in a dispute over a mule. However, it is the legend of the silver bell that endures, a host of tales with varying scripts that claim the bell was buried nearby. For years treasure hunters have riddled the site with potholes, searching for a fortune.

At my last visit, only nine of the original columns remained. The most impressive relic still standing is the fifty-foot-tall chimney of the cotton gin now closely surrounded by trees even taller. The real treasure though is the eighty thousand acres of swamp now protected as Tensas River National Wildlife Refuge. If there is a tolling bell to be heard, it is the ageless melody of birdsong.

Noms de Bayou

Lest you think French influence on our state is restricted to the southern half, consider the sinuous streams of northeast Louisiana.

They flow through our geography with Franco-laden labels both pure and bastardized—and with good reason. Spaniards were the first Europeans to pass through this area. They were transient and too busy searching for gold and incidentally destroying endemic cultures to bother with naming wilderness features. If they did, they did not stick. Frenchmen were the first to establish a lasting presence. They were not Acadians. Throughout the eighteenth century, free spirits floated down from central Canada to trap furbearers and render pots of oil from oleaginous bear carcasses. A handful of "pure" French dragged oxcarts up from the Point Coupee region. Most were not interested in settlement, much to the chagrin of colonial authorities who yearned for the stability of domesticated farmers with pedigreed wives and watertight roofs. The Frenchmen were not on the landscape but of it, like the native women with whom they produced a generation of dark-eyed children. Home was wherever they needed to be to reap the seasonal harvests. Game for the table was available all year. Wild fruits began with mayhaws and dewberries and ended with muscadines and hickory nuts. Fishing was best during the spring overflows and later in the summer when the bayous slowed to wading trickles. Then V-shaped barriers of wooden stakes set across a stream would herd catfish and buffalo into willow-basket traps. Autumn and winter were the times to gather prime pelts from deer, beavers, and otter. Canebrakes were fired to expose bears, and waterfowl borne on Arctic winds were plucked from the cornucopia of natural resources. All of these activities had a common thread. They were on or near waterways. One does not efficiently transport burdens overland through virgin swamps. As had been the case for thousands of years here, dugout canoes of red-heart cypress were the conveyance of the day. Frenchmen paddled the eddies and drifted the currents with thousand-pound bundles of furs, with Indian wives and half-breed children, with apprehension of losing these freedoms. They plied every major stream in northeast Louisiana. They put their names on nearly all of them.

Most names fall in one of two categories: a French surname, or reference to a natural feature associated with the stream. Surname examples include "Bayou D'Arbonne," believed to be derived from the common French-Canadian surname "Derbanne." A Gaspard Derbanne was known to be a companion of St. Denis in his early eighteenth-century exploration of the Ouachita Valley. Galion Bayou in Morehouse Parish was named for a prominent hunter/ trader in that area. "Chauvin" was a surname that yielded Chauvin Bayou and Chauvin Swamp in Ouachita Parish. Bayou Macon comes from the Maconce family. Others in this category may include Bayou Desiard, Bayou Bartholomew, and Choudrant Bayou.

The second category is descriptive. Bayou Lafourche is interpreted as "Forked Bayou," Bayou Coulee becomes "Flowing Bayou," and Bayou de Glaize "Cold Bayou" or "Ice Bayou." Bayou de Butte was named for Indian mounds long since vanished from its shores. Chemin-A-Haut Bayou translates to "High Road Bayou," a reference to the north-south Indian trace that once ran along its flood-free high bank. Plants and animals are represented also. Cheniere Creek refers to the adjacent oak forests. Lapine Bayou is "Rabbit Bayou" and Bayou de l'Outre is "Bayou of the Otter." Imagine bison thundering across the shallows of Boeuf, that is, "Buffalo" River.

Then there are mystery names with veiled hints of an instant of humanity that flowed away with time. What was the "good idea" of Bayou Bonne Idee? Even the two large streams in northeast Louisiana with Native American names, the Ouachita and Tensas rivers, likely have French spellings.

Other geographic features have French names (for example, the Prairies des Canot, Mer Rouge, and Chattlerault), but none embellish our maps like bayous, creeks and rivers. When eighteenth-century Frenchmen plied these streams, they could not comprehend that within two hundred years dams, dredges, and levees would make the water bodies unrecognizable to them, or that relics of their culture would linger with names first spoken from the bow of a pirogue.

[This article first appeared in the 1999 *Louisiana Folklife Festival* publication.]

Roosevelt's Louisiana Bear

In addition to his presidential legacy, Theodore "Teddy" Roosevelt (1858–1919) is remembered as a staunch conservationist, outdoorsman, and hunter. During his second term as president he, along with a group of state and national dignitaries, gathered in East Carroll and Madison parishes in October 1907 for a traditional bear hunt. As revealed in his writings, Roosevelt considered the event his great adventure "In the Louisiana Canebrakes." The Louisiana hunt was Roosevelt's second attempt to bag a southern bear after an unsuccessful 1902 endeavor in Sharkey County, Mississippi, where the legend of the "Teddy Bear" originated.

Prominent Louisiana businessmen John M. Parker and John A. McIlhenny hosted the president. Parker would become Roosevelt's vice-presidential running mate in a failed third bid for the presidency, and later governor of Louisiana. McIlhenny was a member of Roosevelt's Rough Riders in the Spanish-American War and

Black Bear

directed the U.S. Civil Service Commission during his tenure in Washington. Other members of the party included Dr. Alexander Lambert, the Roosevelt family physician, and Presley M. Rixey, the surgeon general of the U.S. Navy. The chief guide for the hunt was Ben Lilly, a legendary backwoods character known for his predator-killing prowess. Bear hounds were provided by two Mississippi planters, Clive and Harley Metcalf. The dogs were tended by Holt Collier, a former slave who was credited with killing more than three thousand bears in his lifetime.

After a brief stop in Lake Providence, the president's special train pulled onto a siding at Stamboul in southern East Carroll Parish about fifteen miles north of Tallulah on October 5, 1907. The party disembarked, mounted horses, and rode west through the uncut forest to a camp on the bank of the Tensas River. Roosevelt was enthralled with the vast virgin swamp and wrote of the giant trees and abundant wildlife. To his son Kermit he penned, "It is a wild country all right, for today across the bayou we suddenly caught a glimpse of two wolves which came down to the water to get a drink." However, the bears did not cooperate. The hunters found bear sign, and the dogs managed to track and jump several, but the president never got a shot. After a week of unrewarding effort except for a deer killed for camp meat, the party moved about fifteen miles south to Bear Lake in Madison Parish. Of this location the president recorded, "Our present camp is in a lovely place, beside a long, narrow lake, or bayou, which is beautiful just at this moment under the moonlight, great cypress trees standing along the bank like sentinels above the rest of the forest."

From the new camp the hunters pursued their quarry persistently. Bears were found by the dogs, but the bruins quickly sought refuge in the impenetrable canebrakes, confounding both hounds and hunters. After several days as the end of the hunt approached, the party decided to change tactics. Instead of the president waiting on a crossing stand for the dogs to drive a bear to him, he would attempt to ride with the pack in hopes of getting a shot at the fleeing animal. The next morning the plan worked. The dogs jumped a bear, and Roosevelt made a desperate, dangerous gallop through the swamp to head them off. The former Rough Rider wrote, "The tough wood horses kept their feet like cats as they leaped logs, plunged through bushes, and dodged in and out among the tree trunks; and we had all we could do to prevent the vines from lifting us out of the saddle, while the thorns tore our hands and faces." After the evening campfire where the day's hunt was rehashed, the president once again wrote Kermit of the adventure, "At last today I killed my big she bear—202 lbs. It was my thirteenth day; now everything is pure pleasure. The dogs had her up for nearly three hours, while we galloped on horseback and ran on foot up and down the outside of the dense canebrakes in which the chase took place, at different times all of us were thrown out. At last one of these planters—very fine fellows—and I heard the bay in a dense canebrake, and slipped in; finally as we crouched motionless the bear walked near enough for me to see her through the cane. She was twenty yards off. I shot her through the lungs at first shot, and then broke her back between the shoulders lest she should injure the leading dog. The hunt was great fun."

Theodore Roosevelt's Louisiana bear hunt was widely reported in the national media. The legacy of the visit often overshadows his more enduring contributions to the conservation of Louisiana's natural resources. He was responsible for creating an assemblage of lands that became the National Wildlife Refuge System. Several of the twenty-four refuges comprising more than 568,000 acres in Louisiana now provide critical habitat for the threatened Louisiana black bear.

[A version of this story first appeared in *KnowLouisiana Encyclopedia,* 2013.]

Sicily Island Neanderthal

On the morning of January 9, 1951, two Baton Rouge newspapers, the *States Times* and *Morning Advocate,* ran a story that fueled coffee-shop gossip and tailgate prattle across the state for weeks to come. The articles described the discovery of "Neanderthal man—an 11 foot tall ancestor of modern man—that lived in North America about 50,000 years ago." The backstory is that two days after Christmas in the preceding year a bulldozer operator was working up a pile of gravel in a pit in the Sicily Island Hills in Catahoula Parish when he noticed what appeared to be bones of a large mammal. They spilled out of the vein of chert about fifteen feet below the natural surface of the ground. By nature a curious man, the operator halted the chugging World War II surplus machine and jumped down from the track to get a better look at the bones. There were a dozen or so fragments in sight, one obviously a backbone, and they certainly looked old. Who knew what others were yet to be revealed in the embankment? With plenty of other gravel to work, the tractor driver decided to move beyond the bone place and work in a different area. Details of what happened next have faded like the old cypress tenant shacks just down the hill from the quarry, but in some fashion the "experts" were notified and a geology professor from Louisiana State University with students in tow showed up to excavate the site. They gathered up leg bones, arm bones, foot and ankle bones, part of a shoulder blade, and five vertebrae. Back at the university the bones were cleaned up a bit, packaged up, and shipped off to the U.S. National Museum for identification. Meanwhile, someone somewhere ascertained the trove to be the remains of a robust-boned Neanderthal, and the newspapers ran with it.

There were, however, problems from the beginning for those who burdened their consciences with facts. Neanderthals lived in Europe and Asia, never in North America. They were not ancestors of modern humans but rather likely a separate species within our genus *Homo*. And the average height of males was almost exactly half the fabled eleven-footers. When the verdict came down from the National Museum, some folks were disappointed. It was only a common black bear, albeit a very old one. As for the newspapers, they were galloping toward new stories with no time to make right old ones.

Signal Trees

In the last few years, GPS devices have become ubiquitous in our culture. Whether one is motoring the maze of big-city freeways or navigating a pirogue through the Atchafalaya Swamp, a GPS unit eliminates all excuses for becoming lost. From a historical perspective, this raises the question of how people navigated across wilderness landscapes two hundred years before Garmin and Magellan. Without a doubt such skills in Native Americans were almost innate because their lives depended on them. One of their techniques observed by early European explorers involved the concept of signal trees. Signal trees, sometimes called trail trees, were intentionally manipulated to grow in a specific shape in order to mark trails, territorial boundaries, shallow fords, sacred places, and the like. Indians made them by bending over a sapling, securing it in place with rawhide or vines, thus forcing the tree to point in a desired direction. Oaks are said to have been a preferred species. Appar-

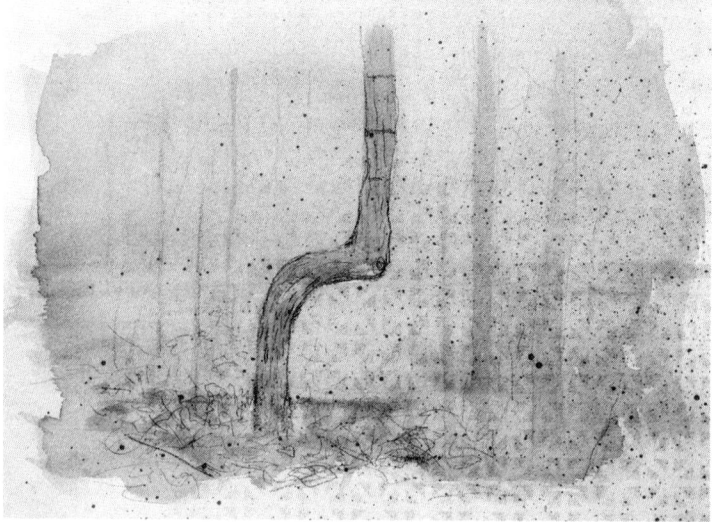

Signal Tree

ently, there were many different configurations of signal trees that depended on the tribe bending them, the geographic area, and what the trees pointed to.

A number of old trees scattered throughout the forests of the eastern United States are considered surviving examples of signal trees based on varying degrees of documentation. Some are marked with commemorative plaques and touted by tourism promoters. Of course, every bent tree in the woods is not a signal tree. Trees often take that shape as a result of wind and ice storms, another tree falling on it from natural causes, or logging activities. A genuine signal tree would likely be at least two hundred years old. For those that are real, imagine the history that has passed within the shade of their boughs.

Sir Henry Stanley

Even before he arrived in the swamps of southeast Arkansas as a young man, he had been around a bit. Born in the north of Wales of uncertain paternity in 1841, he was baptized as John Rowlands. The name didn't stick. He arrived in the port of New Orleans at age eighteen, leaving behind a harsh upbringing in his native country. After a bit of bayou prowling, he was soon traveling the Mississippi Valley as the protégé of a wealthy cotton broker and his adopted namesake. At some point he concluded that a more stable vocation was in his best interest, and he became an apprentice to a German shopkeeper in the frontier village of Cypress Bend, Arkansas. Cypress Bend was a rough-cut steamboat stop just across the big muddy river from Greenville, Mississippi. Life there was filled with environmental hazards, not the least of which was swamp fever or malaria, as we know it today. The young man contracted the disease and suffered its intermittent fevers and chills. A young, familiar woman proved to be a greater peril, though, using feminine social skills of the day to pressure him into a hasty enlistment in the Confederate Army at the outbreak of the Civil War. He joined the Dixie Grays of the Sixth Arkansas Infantry and soon found himself a prisoner of war near Chicago after being captured at Shiloh. After the war he became a journalist, which led him to other swamps on other continents. In addition to malaria, he faced the fevers born of tsetse flies. Instead of cottonmouths and alligators, he avoided crocodiles and vipers. The meanderings of the Mississippi River were well known to men of European descent when he floated upon it. He was the first to trace the course of the Congo. History recalls him as Sir Henry Morton Stanley. Most of us know the man who spent a formative period of his life in southern swamps by a question he once asked of a Scottish missionary on the shore of Lake Tanganyika: "Dr. Livingstone, I presume?"

Groundhog Sawmills

Except for coastal marshlands and tallgrass prairies of the Southwest, Louisiana was historically a world of forests. Virgin stands of longleaf pine in the central part of the state, primeval bottomland hardwoods and cypress swamps, along with upland hardwoods and pines in the hill country were viewed by settlers as both daunting obstacles and coveted natural resources in the form of potential wood products. Logging on a scale that depleted the state's virgin forests, however, did not begin in earnest until about 1900 when reliable steam-powered and later diesel or gasoline sawmills became common. Burgeoning railroads to move logs and products contributed to the elimination of our original forests. By World War II, Louisiana's virgin forests were practically gone.

The sawmills were not all large, industrial-scale facilities. Almost every town or village had a mill of some kind, often called a "groundhog mill." Other than the stereotypical operation that produced lumber from large logs, specialty mills abounded. Shingle mills made roofing shingles, usually from cypress heartwood. Stave mills sawed thin boards that were used in barrel making before steel barrels became common. The best and most valuable barrel staves were made from white oak. Hoop mills cut elm logs into thin strips that were then steamed and shaped into circular rolls.

The hoops were used to hold together many types of wooden barrels, kegs, butter churns, and containers for hoop cheese. Tie mills specialized in sawing crossties for the railroads. They could utilize smaller trees that were not otherwise merchantable. Handle mills were found where hickory trees were abundant. The hard, durable wood was sawn into bolts to dry and later finished into a variety of tool handles. Planer mills smoothed rough-cut lumber into finished products.

In the delta lands most of the original forest was replaced by row-crop agriculture. The same thing happened on higher ground, but the produce was sterile stands of genetically modified pine trees. Both scenarios resulted in the demise of small sawmills and biological diversity, thus ending a short chapter in our cultural history and a long one in our natural history.

Steamboat Disasters

Robert Fulton's steamboat is arguably the single most important invention that spawned settlement and economic expansion in nineteenth-century Louisiana. On a landscape lacking roads but braided with bayous and rivers, travel via water was the only efficient means of transportation. By eliminating the manpower required to row or paddle, often against powerful currents, steamboats fueled an exponential growth in trade and development. However, they were not without hazards as high-pressure steam boilers manufactured in the science of the day were analogous to kegs of dynamite. Steamboat explosions were dramatic, deadly, and common. In the thirty years prior to the Civil War, several thousand lives were lost in steamboat calamities. The disaster of the *Princess* near Baton Rouge in 1859 was a tragically typical example.

In 1859 the *Princess* was a four-year-old state-of-the-art side-wheel paddleboat. A sister boat to the famous *Natchez,* she had undergone a thorough retrofitting the previous summer and was said to be one of the fastest and most luxurious crafts on the Mississippi River. Lavish meals were served four times a day in a great central hall, and surviving menus list such gourmet delicacies as broiled pompano and stuffed crabs. Her clientele were among society's elite in the Lower Mississippi Valley. The *Princess* ran weekly round trips from New Orleans to Vicksburg and back, departing the New Orleans wharf promptly at 5:00 p.m. every Tuesday. On the three-hundred-mile upriver leg she made stops at Donaldson, Plaquemine, Baton Rouge, Port Hudson, Bayou Sara, Red River Landing, Fort Adams, Natchez, Waterproof, Rodney, St. Joseph, Grand Gulf, and Warrenton before arriving at Vicksburg. The stops were reversed on the downstream journey as passengers, mail, and tons of freight, including four-hundred-pound bales of cotton, were loaded and unloaded.

Maintaining a posted schedule was essential in the competitive business of steamboat commerce. When the *Princess* pulled up to the wharf in Baton Rouge early on the morning of February 27, 1859, she was already late. The boat was loaded with passengers, mostly from Mississippi and Louisiana, headed to New Orleans to celebrate Mardi Gras. More passengers boarded at Baton Rouge, including a number of politicians fresh from the state legislative session that had just ended early for the holiday. An estimated four hundred people were on board the *Princess* when she pulled out into the current of the river after 9 a.m. Because the boat was late, high boiler pressure had been maintained during the stop, and second engineer Peter Hersey was reported to have declared that he would make it to New Orleans on time "if he had to blow her up." As a portent of the looming catastrophe, the Mississippi River was veiled in a dense fog.

The *Princess* was about six miles below Baton Rouge at Conrad's Point when a teenage boy watching the boat glide along from a distance noted, "a great column of white smoke suddenly went up from her and she burst into flames." The explosion was cataclysmic as all four huge boilers burst at once. Peter Hersey and many others died instantly in a blast of scalding steam. Concussion swept away the infrastructure, and the upper cabins, state rooms, and hurricane deck collapsed inwards. Fire broke out and began to consume the remains.

Even amid the horrendous chaos, rescue efforts began immediately. The flaming hull drifted onto a shoreline sandbar and grounded. Intact crewmen and passengers dragged the injured up onto the sandbar. Slaves from the nearby Cottage plantation were ordered to bring sheets and blankets. Barrels of flour were emptied on the ground, and the terribly burned victims were rolled in it and placed in the shade. Men in skiffs from both riverbanks rescued people clinging to debris. Plowing upriver from New Orleans, the *Natchez* was the first steamboat to arrive on the scene. For several hours her crew and passengers provided aid before heading upriver with her decks covered with bodies of the dead and injured.

There was no manifest to record the names of passengers aboard the *Princess* at the time of the disaster. The number of people killed instantly or who drowned or died as a result of their injuries was variously estimated from seventy to two hundred; the actual number was likely closer to the smaller figure. At least a hundred people survived their injuries. Among those killed were Louisiana state representatives H. J. Huard and Charles Bannister. The boat and its entire cargo were a total loss. Human error—failure to maintain safe boiler pressure—was determined to be the cause of the tragedy, and a pall was cast over the 1859 Mardi Gras celebrations.

[A version of this story first appeared in *KnowLouisiana Encyclopedia,* 2014.]

Wash Pots

In retirement they seem innocent enough, often sitting quietly in the side-yard holding bouquets of pansies. Back in their day though they were instruments of hard manual labor, especially for Louisiana women who dreaded their weekly encounters. For them, cast-iron wash pots were undesirable necessities.

A review of estate inventories during legal successions reveals that wash pots may have been one of the most commonly found items in nineteenth-century Bayou State households. The typical pot was about eighteen inches in diameter with a rounded bottom. It held about twenty gallons of water. Three short, stubby legs when placed on rocks balanced the kettle, and a pair of opposing iron loops on the rim could support the pot if hung by a chain. The exterior was always charred sooty black from the fires that heated the contents.

It was called a wash pot for obvious reasons. More than anything else it was used to wash clothes. Wash day was a laborious ritual that involved building and maintaining a hot fire under the pot so that clothes could be boiled before undergoing a series of rinses in nearby tin tubs. Accessible water was a necessity as each washing required several pots of heated water. Near my house local women once gathered at a clear, flowing spring to wash.

Wash pots had other uses as well. At hog-killing time water was boiled in the pot and poured into a drum in which a freshly slaughtered hog was immersed. The hot water sterilized the pig and loosened the hair for removal. When the hog was butchered, the scraps were thrown into the wash pot to boil, rendering the fat and yielding cracklings that floated to the surface. Sometimes a small can of lye was added to the fat, and the result was lye soap, another household necessity.

By the 1930s mechanical washing machines became common and the prevalence of cast-iron wash pots began to diminish. History has not recorded one incident of a Louisiana woman lamenting this occasion.

Hunting

No one would argue whether hunting and fishing are popular pastimes in Louisiana. Our state harbors healthy populations of many types of animals and fish considered game species that are pursued by thousands of hunters and anglers. Because of the high public demand to engage in these activities, a complex and constantly evolving set of regulatory laws is necessary to protect and provide for a sustainable use of the natural resources. Consider that the latest issue of the booklet describing state hunting regulations comprises more than fifty pages, and this does not include federal migratory bird hunting or fishing regulations.

I recently came across a comparable listing of Louisiana game and fish laws that was issued eighty-five years ago. Life was simpler then. All of the state hunting, fishing, and trapping rules were compiled on three pages of a small brochure. Hunting and fishing licenses cost one dollar each. It is interesting that some regulations were less restrictive then and some more so. The limit on ducks was twenty-five daily and three hundred in a season. As a retired federal game warden, I can't imagine how the season limit was enforced. A hunter could also kill five bears in a season. Beavers and wood ducks were completely protected, but bobcats, cougars, vultures, hawks, and owls could be killed on sight. Fishing season was closed statewide from March 1 to April 30.

Some of these regulations seem asinine now, and some were very harmful to wildlife populations. It is important to remember though that, then as now, fish and wildlife laws are a reflection of the current biological knowledge of the species and the political persuasions of the day. Biological wisdom has greatly increased in the last eighty-five years; political acumen not so much.

1835 Balloon Accident

As with other cultures around the world, the first human inhabitants of what is now Louisiana undoubtedly marveled at the mysterious ability of birds to fly and perhaps even yearned to soar over the bayous and swamps free from the bonds of gravity. They could not imagine a night sky in which the blinking of aircraft lights or satellites was always present somewhere in the heavens as is the case today. Humans of course have mastered flight through a series of technological advances but not without peril even when Louisiana was a fledgling state. The following newspaper account reveals an 1835 Icarus-like episode involving a hot air balloon.

Mr. Elliott, the aeronaut, has attempted to make an ascension in New Orleans, but the wind proved to be too strong. After seating himself in his balloon, and cutting loose, he was swept violently across the arena, knocking down several persons in his passage. The balloon next encountered a chimney top, which was overthrown by the concussion, and Mr. Elliott's thigh was broken. Part of the bricks of the chimney falling into the car, prevented the balloon from rising higher, and it was afterwards dragged over housetops and walls, and dashed against windows, till the aeronaut's hands, face and head were shockingly cut and mangled. At length, the cords of the balloon became entangled on the masts of two vessels in the river, and fortunately for Mr. Elliott, his farther flight was checked. In his passage over the buildings in the city, some of the cords by which the car was attached to the balloons, were sundered, and the aeronaut afterwards smiled with his head nearly downwards. If he recovers from his wounds and bruises, he will owe his life mainly to the great presence of mind that he maintained amid all the perils through which he passed.

I suspect that when the dust of this affair settled Mr. Elliott had an even greater appreciation of birds.

The Bear and the Silver Spoon

President Theodore Roosevelt was frustrated when he arrived in East Carroll Parish in October 1907. An avid hunter, he had long desired to kill a black bear on a traditional southern hunt with baying hounds and moss-draped swamps as a backdrop. His first effort in Sharkey County, Mississippi, five years earlier had been unsuccessful except for spawning the iconic "Teddy Bear" stuffed toys when he refused to shoot a young bear that had been tied to a tree by his hunting guide. Now after a week of unrewarded effort in East Carroll Parish, Roosevelt began to grumble that maybe they were in the wrong location. His infamous guide, Ben Lilly, agreed and the hunters moved about fifteen miles south to Bear Lake in Madison Parish. There on the thirteenth day of the hunt the dogs bayed in a dense canebrake and the president got his trophy.

After a day of rest the hunters broke camp, and Roosevelt rode back to his waiting train. On the way he spent the night at a nearby plantation managed by Mr. and Mrs. Leo Shields. During supper that night Mr. Shields mentioned the need for a local post office as they were then dependent on mail via Mississippi River steamboats. The next morning Roosevelt departed on the train, and for a short way the Shields' two-year-old daughter rode on the president's knee. Soon after the visit a post office was authorized for the area and promptly named "Roosevelt" by the local people. A historical marker on a barren stretch of U.S. Highway 65 is a lone reminder of the event.

In 1987 my wife, Amy, interviewed the Shields' daughter, who was then eighty-two years old and known fondly as Poche or Mrs. W. Z. Adams. She was of course too young to remember the story when it happened, but her family related it to her many times when she got older—and she had hard evidence. Several months after the president's visit, Poche received in the mail a silver spoon embellished with a Teddy Bear. Amy held the spoon and noted that it was inscribed "To little Mip Agnes Tabitha Shields with all good wishes for her future from Theodore Roosevelt, October 21, 1907." According to Poche, "Mip" was a term of affection used for small girls.

In hindsight, had the bear been a bit faster, Roosevelt might not have been as amenable in doling out his post office and silverware in northeast Louisiana.

Poverty Point

World Heritage sites are places deemed by the United Nations to have cultural or natural significance on a global scale. Poverty Point, a prehistoric cultural site of exceptional merit in West Carroll Parish, was recently added to the sparse list of those in the United States that includes the likes of Grand Canyon and Yellowstone national parks. Perhaps the most amazing thing about Poverty Point though is how little we know about it and the curious people

Poverty Point

who lived there. Unsolved mysteries drift along the shore of Bayou Macon like silken threads of ballooning spiders. My thoughts on the matter lead me to ponder that:

- This hallowed place is *not* Poverty Point. It has a name that we cannot know, cannot imagine, perhaps that our modern tongues cannot pronounce. It is a name that will never be spoken again.

- The people of this place shaped their world with switch-cane baskets, stardust, and a cosmic blueprint beyond our comprehension. Six mounds, six ridges—enigmatic architecture that makes profane strip malls and fast-food joints.

- Their larders were forests of persimmon, pawpaw, and pecan; rivers of catfish, buffalo, and drum. They cooked with fired earth and left the swirls of their fingerprints for our imaginations to misinterpret.

• In their mind's eye from atop the big mound even the old people with cataracts could see five hundred miles to northern quarries of galena, flint, quartz, and soapstone. Beckoned by shamans, treasure from these places drifted downstream to the ridges.

• Ancient human bones are nowhere to be found on the site. Did the small clay heads speak of burial taboos? Should we know that the tiny red owls called for inherent sacredness in such a setting?

• Before the birth of this place, before the flowering of earthworks, 260 generations flowed across this landscape, absorbing the wisdom of time in place. Along the same bayous we are as newborns in all that matters.

Louisiana Bison

The image of thundering herds of buffalo racing across endless prairies is not one that is often associated with Louisiana, the Bayou State. Historically, though, the scene is not far-fetched.

The animals we call "buffalo" are more correctly termed "bison" to separate them from true buffalo of Africa and Asia. Early French explorers in Louisiana called them *boeuf sauvage*—"wild ox." Formidable in appearance, the bulls stand six feet high at the pronounced shoulder hump and weigh as much as a ton. Both sexes have a massive head, neck, and shoulders and are robed in a thick, wooly pelage. They were the largest land mammal to inhabit Louisiana in historic times.

That they did indeed live in Louisiana is well documented. Our maps denote three Bayou Boeufs, Boeuf River, and Boeuf Lake. As part of the continental "southern herd," they ranged across most of the state at least part of the year but not likely in the tremendous numbers associated with those of western and northern prairies. Bienville reported killing a bison near what is now Winnsboro in 1700. Penicaut wrote of shooting twenty-three bison at Bayou Manchac in 1712. Other eighteenth-century accounts mentioned bison near present-day Baton Rouge and New Orleans. By 1800, bison seem to have been almost eliminated from the state's list of magnificent fauna. One historian wrote, "The last buffalo seen in the neighborhood of Fort Miro [now Monroe] was killed in 1803." This pattern continued for the next hundred years until the entire continental population, estimated at sixty million, was market-hunted to near extinction. Bison are found today in Louisiana in a few small captive herds scattered around the state.

One of the biggest surprises of my life involving wildlife occurred a few years ago in a remote marsh in southwest Louisiana. While hiking along a low ridge at dusk I could hardly believe my eyes as a buffalo emerged from a wax-myrtle thicket. She was soon followed by two more, and they began grazing as darkness fell over the vast, fenceless marsh. Later I learned that they appeared in this area soon after Hurricane Rita and were thought to have come from Texas during the storm. For a moment though I was in the eighteenth century.

Shooting Stars

Of all natural phenomena, one that never fails to elicit a cry of exclamation is a bright shooting star. Who among us has watched a meteor streaking across the night sky and remained silent? Meteors occur when dust left in the wake of passing comets enters earth's atmosphere and burns up. When concentrated swarms of dust are encountered, the result is a meteor shower with up to one hundred meteors per hour. Each year the earth crosses several comet-dust trails at specific times during its orbit around the sun. The Lyrid meteor shower occurs April 21–22, the Perseids on August 11–12, the Leonids on November 17–18, and so forth. The names of the events refer to the constellations from which, due to an optical illusion, the meteors appear to originate. Some of the littering comets though long gone can even be identified. Thus, the Perseids appear to be associated with comet Swift-Tuttle and the Orionids with comet Halley.

Perhaps the most spectacular meteor storm in recorded history occurred on November 13, 1833. During a four-hour period beginning at midnight, the skies were lit by thousands of shooting stars each minute. Every person in North America was likely aware of the event. This was a time before electricity and light pollution when nights were truly dark. The nocturnal flashes of light would have been so bright and unusual as to awaken even sleeping people. Newspapers across the country wrote of the incident. People thought the world was coming to an end, that the whole heavens were afire, that all the stars were falling. A preacher in Virginia wrote that sinners believed the Judgment Day at hand and fell on their knees in penitence. Sioux Indians recorded the event in their sacred winter counts. Louisiana was not spared the episode. Dr. R. F. McGuire, a Monroe doctor and lawyer, recorded in his diary: "A most singular rain of fire from the heavens appeared to start from a center almost east, at an elevation of about 55 degrees above the horizon, creating a light like daybreak and every few minutes one so large as to produce a glare. Thousands in view at once, from 12 o'clock until daylight, the most grand display of fireworks ever witnessed." We should be so lucky!

Bayou Topogeny

There is little doubt that most indigenous cultures perceived their connectivity to and place on the landscape very differently from the conceptions of modern society. The relationship was knowledgeable to the point of intimacy with the geography of their lives. Such wisdom was and in many cases remains critical for survival. Knowing and labeling the places where fish concentrated, where the enemy lurked, where refuge could be found in the time of flood was essential. Anthropologists have noted that some groups developed an ordered procession of place names similar to a succession of ancestral names or genealogy. Each list is termed a "topogeny." Oral recitation of topogenies was a way to maintain and pass along critical information to successive generations.

We have no way of knowing what a translated, Native American topogeny along a Louisiana bayou might have sounded like. Surely it would have morphed during the thousands of years of prehistoric occupation. We can, however, for the sake of example create a historic topogeny. For Bayou D'Arbonne, beginning at its confluence with the Ouachita River and paddling twenty-seven miles upstream to the dam, the list could include:

> Mouth of Bayou D'Arbonne → White's Ferry → Dry Slough → Nelson's Slough → Wreck of *Rosa B* → Long Reach → Catfish Slough → Gum Cut-off → Cook's Wood Yard → Whiskey Still Slough → Ducote Slough → Cross Bayou → Boggy Bayou → Little Choudrant Bayou → Willow Hole → Big Choudrant Bayou → Steamboat Hole → Holland's Bluff → Long Slough → Eagle Lake → Old Mill → Lake Drain → The Wreck (of the *Tributary*) → Rocky Branch → Turkey Bluff → Meek's Landing → Francis Creek → Rugg's Bluff → Lake D'Arbonne Dam

Some of the listed features are obvious while others, like the steamboat wreck sites, are impossible to discern without prior knowledge of events that occurred many years ago and that was passed along by word of mouth. Similar elements would likely have existed in a prehistoric topogeny of the same bayou.

Bowie on the Bayous

The name "Jim Bowie" often evokes images of a large, fierce hunting knife and a desperate battle for Texas freedom at the Alamo. Although branded a hero in Texas history, Bowie spent most of his life along Louisiana bayous, where records affirm that his aspirations were routinely pursued with less than honorable behavior.

James Bowie was born in 1796 in Logan County, Kentucky. In 1803, while Thomas Jefferson was working out the details of the

Louisiana Purchase, Bowie's father obtained a Spanish grant of eight hundred arpents along Bushley Bayou in Catahoula Parish. There in a wilderness setting Jim spent much of his early boyhood. The family moved once again in 1809 to St. Landry Parish near Opelousas. In 1815, nineteen-year-old Jim struck out on his own along Bayou Boeuf in Avoyelles Parish, where he purchased land and slaves on credit and began cutting virgin timber and floating it to downstream markets.

Jim Bowie's adult behavior revealed an ambitious opportunist who did not permit matters of honesty and moral conduct to stand in the way of personal gain. Congress had abolished the African slave trade in 1808, but expanding agriculture in the Deep South resulted in a greater demand for labor than could be met with domestic slaves. The consequence was a surge in slave runners, including the mercurial French privateer Jean Lafitte. From headquarters on the Texas coast just west of the Sabine River, Lafitte sold his pirated contraband to Bowie, who devised a plan to smuggle them into the Louisiana interior. He then claimed to have captured the illegals and turned them over to authorities for a reward. As per the law, officials then sold the slaves at auction, and Bowie bought them back for resale—this time with a legal title. Dozens of slaves were involved and considerable profits accrued by Bowie.

Bowie's most ambitious ploys stemmed from the chaotic state of Spanish land grants and land titles following the Louisiana Purchase. Compounding the problem, most of the Spanish records had been moved out of the country. Bowie saw an opening and began to personally forge Spanish land grants of prime bayou-front properties in several parts of the state. He then boldly manufactured deeds of sale of the grants to himself. The scale of the ruse was astounding as he claimed up to eighty thousand acres in Louisiana and almost as much again in Arkansas. Bowie's claims were immediately suspect when he attempted to formally register them, but his conniving and political influences kept the matter alive throughout the 1820s. He was even able to sell some of the counterfeit titles and reap a profit before the eventual collapse of the scheme.

Jim Bowie's unscrupulous business practices gained him many enemies in a culture where a slight often resulted in a deadly duel. One disagreement involving Rapides Parish sheriff and banker Norris Wright resulted in Wright shooting Bowie point-blank with a pistol. Poorly armed at the time, Bowie survived the deflected shot but vowed to never again be without a large knife in his belt as long as he lived. Accordingly, the legend of the Bowie knife was born, and the stage was set for a gruesome encounter. On September 19, 1827, on a Mississippi River sandbar near Natchez, Bowie was present at a duel between Dr. Thomas Maddox and Samuel Wells III. No one was injured when the principals exchanged shots. The affair seemed over until antagonism between members of the principals' entourages flared into a melee. Alexander Crain shot General Samuel Cuny. Wright shot Bowie through the lower chest. George McWhorter shot Wright in the side, causing a flesh wound. Bowie drew his famous knife and attempted to chase Wright until he was shot in the thigh by another gunman. Wright and Alfred Blanchard stabbed Bowie with sword canes. In a desperate lunge, Bowie grasped Wright by the collar and thrust his long knife into his enemy's chest, killing him instantly. The violence ended abruptly, and attending physicians rushed to treat the injured. Cuny and Wright were dead, and Bowie's recovery took months. A grand jury was convened afterward but handed down no indictments. The brawl made national news and enhanced Bowie's notoriety.

By the end of the 1820s, Bowie's land schemes were crumbling on all fronts, and there was an increasing chance that he would be held legally accountable. Having made several brief trips to Texas in recent years, Bowie sensed new opportunities there and reprieve from his longstanding troubles. In early 1831 he sold most of his remaining assets and moved to the tumultuous territory of Texas. Once again he became involved in land schemes and shifting politics. In that vein, his life ended on March 6, 1836, defending the Alamo in the name of freedom for a new republic of Texas. In spite of his past life, most of which was spent along Louisiana bayous, Jim Bowie's final hours ensured his legacy as a hero.

[A version of this article first appeared in *KnowLouisiana Encyclopedia,* 2013.]

Wild Women

In the twentieth-century conservation arena, battles were tumultuous and dominated by males. Great legal strides were made in protecting our country's vital air, water, and other natural resources, often in opposition to the wishes of powerful corporations and politicians. Paradoxically, three of the most potent warriors on the side of conservation were small-framed, soft-spoken, selfless people. Their weapons were ecological wisdom, persistence, and courage. They were also women.

Rachel Louise Carson was a career marine biologist for the U.S. Fish and Wildlife Service. A well-known natural history writer, she became a social critic with the publication of her book *Silent Spring* in 1962. The book revealed the damaging effects of pesticides on the environment, especially DDT and its terrible impacts on birds by causing eggshell thinning. Carson accused the chemical industry of covering up the truth about their products. *Silent Spring* spurred an overhaul of national pesticide policy and is often credited with launching the global environmental movement. For her work Carson was awarded the Presidential Medal of Freedom.

Marjory Stoneman Douglas's mission was to save the Everglades from drainage and development. During many years of writing, speaking, and lobbying, she elevated the plight of the Everglades on the national agenda, which resulted in the preservation of most of that vast ecosystem. Her book, *The Everglades: River of Grass,* was instrumental in the struggle. Former assistant secretary of the interior Nathaniel Reed described Douglas as "that tiny, slim, perfectly dressed, utterly ferocious grande dame who can make a redneck shake in his boots." Douglas also received the Presidential Medal of Freedom, at age 103.

Closer to home, Caroline Dormon dedicated her life to conservation in the Kisatchie Hills of north Louisiana. Her love of native plants and all things wild resulted in expertise in the fields of forestry and horticulture. She was also an accomplished botanical artist and author of six books, and as a lecturer in demand she traveled widely. A persistent lobbyist, Dormon was instrumental in the establishment of Kisatchie National Forest in spite of well-heeled logging interests. Her legacy lives on in this forest encompassing more than 600,000 acres and in the Caroline Dormon Nature Preserve in Natchitoches Parish.

Beneath the meek countenances of these three women, compelling spirits and lion hearts prevailed to make our world a better place to live. Especially in these days, we need many more of their mettle.

Bayou/Human History

When I am prowling about local rivers and bayous, I often contemplate those who came before me and wonder about their life experiences. Some Louisiana history books tend to begin serious discussion of this region with the arrival of European explorers and later settlers. Starting with Hernando de Soto's rampage across the Southeast in the mid-sixteenth century, the first two hundred years of European presence were mostly just flickers of occurrences by Spaniards and Frenchmen looking for treasure in the form of gold or animal pelts. Gold was not to be found, but furbearers were abundant enough to eventually attract a growing contingent of French Canadian trappers. Of their race, they were the first to embrace the four-hundred-plus bayous in Louisiana as major transportation arteries. On the surface they were low-impact visitors leaving behind few visible signs on the landscape to mark their passing. Their invisible traces though, in the form of Old World pathogens, were not so innocuous.

Settlers with shifting allegiances according to the politics of the day followed to sink foreign roots into native soils along the sinuous bayou banks. With the Europeans' steel plows and domestic livestock, widespread environmental alteration began at an unprecedented pace. For the next 250 years until today the newcomers continued to change the natural world forever. The newcomers are all of us—descendants of Europeans and Africans who migrated, many involuntarily, to what has become the Bayou State.

It would be easy to consider this information an appropriate summary of humans and bayous. Textbook writers have done so and created a historical distortion. Europeans and Africans were latecomers on this landscape—very late. To portray the scale of imbalance, consider that native people lived along the bayous for four hundred generations before the Johnnies-come-lately arrived. After contact, the waters of our cultures roiled with conflict for four generations until the natives of this region evaporated as a distinct race. As late arrivals we have now been present about nine generations. Think about that: four hundred generations of Native American occupation versus nine generations for us immigrants. Along the bayous we are as newborns. What history has been lost?

Louisiana Ferries

Louisiana's bayous and rivers have long been considered blessings and banes, depending on one's preferred mode of transportation. In a land laced with aquatic arteries, streams were the only practical means of conveyance for centuries. Only when colonial authorities began planning a system of roads to facilitate European settlement and economic development did the waterways become appreciated as substantial barriers to progress. To overcome the aquatic obstacles, bridges were built when feasible, but often the streams were too large or the water level fluctuations too drastic to be surmounted by the engineering of the day. As a result, the problem was addressed with ferries—boats that transported people and their property across otherwise impassable water bodies.

The issue was of such importance that one of the first laws enacted by the U.S. Territory of Orleans (which later became the state of Louisiana) in 1805 stated that "the judge of every county within this territory may and shall grant licenses for keeping ferries." The authority was later granted to parish-police juries. The records of newly organized Union Parish are typical. In June 1839, William Wilkerson was given a license for one year to operate a public ferry on Bayou D'Arbonne at the mouth of Corney Creek. He was allowed to charge $1.00 for wagons with four wheels, $0.50 for wagons with two wheels, $0.25 for a person with a horse, $.12½ for a person on foot, and $.6¼ for all other animals. The following month, ferries were authorized on Bayou De L'Outre and the Ouachita River. In 1840 the procedures were amended to sell the ferry operations to the highest bidder and to reduce the rates by half.

By the second half of the nineteenth century, dozens of small ferries were operating throughout the state. Most of the boats were decked barges built of heart cypress for durability. Often fastened to a rope or cable stretched between the two banks, they were poled or pulled across by the ferryman, who took advantage of favorable currents when available. Some had ramps at each end that could be adjusted to facilitate boarding and unloading. When resources were available, ferries adapted to new technology. As early as 1823 a steam-powered ferry was operating on the Mississippi River in New Orleans. Eventually all of the larger ferries were gasoline or diesel powered and had hulls built of steel.

One by one, bridges sank the ferries. Vehicles increased in size and number, overwhelming the capabilities of small boats. New gasoline taxes provided revenue to build modern steel bridges. Today, only five public ferries operate in Louisiana. Three are on the Mississippi River, and one crosses the Calcasieu River and ship channel. All large, modern vessels that can carry dozens of vehicles, they are located in places where bridges aren't practical. The least known public ferry crosses the Ouachita River in a remote section of Catahoula Parish with a capacity of six cars per trip. Known as the Duty Ferry, the small boat retains a bit of the ambience of its once-common riverine ancestors.

World-Record Alligator—A Second Look

When humans ponder alligators, size matters; consider the news-worthiness of the discovery of a giant alligator, even in areas where the reptile is common. Throughout the recorded history of American alligators, accounts of colossal alligators are abundant and often accepted as truthful in spite of a lack of hard data or evidence to affirm such reports. Make no mistake, within the biological potential of the species, very large alligators have been scientifically documented; the widespread, well-monitored harvests in recent years have yielded extraordinary individuals. Suspiciously, none of these documented cases remotely approach century-old claims. A closer look at circumstances surrounding the report of what is usually considered the largest alligator on record is interesting and revealing.

Edward Avery McIlhenny was called Ned by his family and friends. He grew up on his father's plantation, which encompassed the ancient salt dome known as Avery Island on the Louisiana coast about twenty-five miles south of Lafayette. The family business included the production of Tabasco-brand pepper sauce on this rare mound of elevated ground surrounded by thousands of acres of pristine wetlands. Natural productivity of virgin marshes in a near-tropical climate is tremendous and was reflected in the abundance and diversity of wildlife that included a cornucopia of seafood and clouds of wintering waterfowl. Alligators were likely as abundant here as anywhere on earth.

According to Ned, he and two assistants departed Avery Island on January 2, 1890, in a lugger and proceeded south through a maze of bayous to Vermilion Bay to hunt geese. Sailing southwest across the bay, they were becalmed at dusk near the mouth of a bayou that had been filled with silt by a hurricane a few years before. The shallow bayou, now cut off from the bay, led to Lake Cock, several hundred yards inland. Ned, seeking game for supper, decided to walk the bank of the old stream with his shotgun. He shot two mallards just before dark, and when wading into the marsh to retrieve them saw what he first thought was a partially submerged log. Approaching it, he discovered that it was a very large alligator that seemed to be addled in the cold air and water. He shot the alligator in the head, presumably with birdshot, and killed it; it was the largest alligator he had ever seen. After spending the night on the boat, Ned and his companions went back the following morning to retrieve the carcass. Using a rope, the three of them tried to pull the alligator through the marsh to the bank in order to skin it. The alligator was so large and the bottom of the marsh so boggy that they could only manage to move the animal a short distance. Ned then decided to abandon this effort and measure the alligator where it was. Using the barrel of his shotgun, which he knew to be thirty inches long, he measured the alligator three times. He then declared the total length to be nineteen feet, two inches.

It is a bit surprising to me that this record is still apparently accepted by many in the scientific community. If a similar claim were to occur today without corroborating evidence, it would almost certainly not pass muster. The fact that the record is based solely on Ned's word is problematic even though he became a well-regarded naturalist and wrote a groundbreaking book titled *The Alligator's Life History* in 1935. Material in his book is obviously based on long-term observations, experiments, measurements, and extensive notes. Ned was likely the most knowledgeable alligator expert in the country at one time. This and especially the fact that he became a prominent, wealthy businessman probably discouraged criticism of a record that otherwise might be questioned.

Issues pertinent to the record include Ned's possible youthful bravado; he was seventeen when he killed the large alligator. Even though he was unable to retrieve the specimen, it would not have been impractical to procure the head, a common tactic of contemporary naturalists. If one assumes that Ned's shotgun barrel was exactly thirty inches long, a standard length for such guns, the alligator would have been exactly 7.66 barrel lengths long—a dimension that in my experience of measuring large alligators in the marsh with steel tapes would be difficult to determine.

A look back into Ned's writings reveals some proven discrepancies. At one time he boasted of first introducing nutria to Louisiana after acquiring the original stock in Argentina. The McIlhenny Company website now states that this claim is untrue. The company historian has said of Ned, "He was well-known on the island for his gift for spinning yarns. . . . I think he saw himself as an entertainer when relating his personal history. He took liberties in a good-natured way." Ned wrote the account of his record in his book published thirty-three years after the incident. With so little evidence, we will never know the truth of Ned's youthful encounter with the giant alligator.

Of course, there are other questionable claims of extremely large alligators. None, though, have been accepted to the degree that McIlhenny's has. Researchers in Florida have investigated many such reports in that state. To verify historical claims, they devised a mathematical model based on a control group of alligators that had been carefully measured. The model demonstrated that the total length of an alligator could be accurately predicted by the known length of its head. The researchers evaluated several old skulls and compared the declared length with the length predicted by the model. The skull of one "record" specimen was measured, and the total length of the entire alligator was estimated to be fourteen feet, ten inches. The early naturalist had claimed the animal was just over sixteen feet long. Researchers found this tendency to exaggerate to be the norm when investigating the old skulls.

Detailed information has been gathered on hundreds of thousands of alligators harvested since intensive management of the species began in the early 1970s. In terms of total length, the contemporary record seems to be a giant that was killed in Alabama in 2014. Biologists there measured him at an even fifteen feet (although greater lengths of this individual were reported by the media).

No doubt, the practice of verification won't tame tall tales. It is as if there is a primal need to believe in the existence of potential human predators on a scale that exaggerates reality. Such ideas lurking around the edges of our domestic imaginations and roaring from the swamps of our psyche seem too much to give up.

[Adapted from my *American Alligator: Ancient Predator in the Modern World* (University Press of Florida, 2013)].

Passenger Pigeons

On November 20, 1869, an entry in a Monroe newspaper noted, "The wild pigeons are making their way southward. We have seen several flocks flying over." Not to be confused with the feral rock pigeons common in city parks and native to Europe, the subject of this article was the passenger pigeon, perhaps the most abundant bird in the world at one time. Observers described their flocks on a scale that seem unbelievable to us today. One flight in southern Canada in 1866 was said to be a mile wide and three hundred miles

Passenger Pigeon

long, and took fourteen hours to pass a single point. The flock was estimated to contain 3.5 billion pigeons. Millions of these birds migrated to Louisiana each fall.

Passenger pigeons were similar in size to the familiar rock pigeons but with a streamlined body shape more like a mourning dove. Acorns, chestnuts, beechnuts, other seeds and berries were the mainstay of their diet. Highly social, they foraged, nested, and roosted in huge communal groups. In a breeding colony a single tree could harbor a hundred nests, and during migration the density of birds alighting for the night would be so great as to break the limbs from large oaks. This collective behavior led to their downfall when the birds were shackled with a monetary value as a source of food and fertilizer.

Although deforestation by pioneers contributed to the decline of passenger pigeons, it was commercial market hunting on a scale that exceeded the slaughter of American bison that brushed the species irretrievably from the face of the planet. Especially vulnerable in breeding colonies, pigeons were shipped by the boxcar-load to eastern cities. From one end of their migration corridor to the other, pigeons were without refuge. A biological tipping point was reached in the 1890s, from which there was no chance of recovery. The last passenger pigeon, a female named Martha, died in a zoo on September 1, 1914. Then as now, elegiac essays about the extinction of a species are only as a feather in the wind.

Louisiana Hippos

When it comes to politics, especially in Louisiana, one really can't make some of this stuff up. Absurd political conduct has a long history in the Bayou State, as illustrated by Congressman Robert Broussard's legislation, H.R. 23261, introduced in the U.S. House of Representatives in 1910. Formally referred to as "A bill to import wild and domestic animals into the United States," the press labeled it the "American Hippo Bill."

The back story is sinuous and involves two unrelated issues of the times, immigration and an invasive species. At the turn of the twentieth century, immigrants were flooding into America and creating a demand for beef that could not be met by farmers. The result was a general meat shortage in the country. A couple of decades earlier, in 1884, New Orleans had hosted the World Cotton Exposition. During this event Japan, as a goodwill gesture, gifted the state a collection of attractive water-hyacinth plants that soon escaped into the maze of south Louisiana waterways and promptly degraded vast wetland ecosystems. Mr. Broussard set out to solve both of these problems with his bill.

The law would authorize the expenditure of $250,000 in federal funds to import wild hippopotamuses from Africa to be released into wetlands of southeastern states. In theory, the animals would gorge on the invasive water hyacinth. When well established, excess hippos could be harvested to address the national meat shortage. Initial public reaction to the proposal was positive. President Theodore Roosevelt thought it a good idea. An encouraging editorial in the *New York Times* referred to hippo meat as "lake cow bacon."

During the Agriculture Committee hearings to consider the bill, a so-called expert testified that hippos could add "1,000,000 tons of meat a year to our supply." Another who had lived in Africa said they were "as tame as a common garden cow." Lengthy testimony involved the taste of hippo meat and the means to control them should they get out of hand. For reasons unknown to me, the bill never became law. It's a good thing for many reasons, not the least of which was that supporters also advocated the introduction of African buffalo and warthogs.

Ben Lilly

At the turn of the twentieth century, Louisiana's vast natural resources in the form of virgin forests and teeming wildlife were besieged by commercial interests and others lacking environmental mores. In this state of diminishing wilderness, Ben Lilly emerged from the swamps of northeastern Louisiana to become a folk hero. His reputation as the best hunter of his day evolved as a result of his obsessive compulsion to kill bears and cougars. President Theodore Roosevelt hired him as his chief guide during his noted Louisiana bear hunt. Ironically, Lilly's successful efforts in Louisiana and later out west contributed to the loss of a lifestyle that he cherished.

Benjamin Vernon Lilly was born in Wilcox County, Alabama, in 1856. As a young man he settled on his uncle's Morehouse Parish farm near Mer Rouge. He hated farming, and dabbled in the cattle and timber businesses. None of these occupations were satisfying. He discovered his passion in the local swamps of Bonne Idee and Boeuf after killing a bear with a knife. From that point forward, his life centered on the pursuit of large predators. Accordingly, in 1901 he transferred his property to his wife and children and walked out of their lives.

Lilly soon learned that he could make a living as a hunter and became good at it. The U.S. Bureau of Biological Survey hired him to collect for the national museum. Over the years he sold them many skulls and skins. From Louisiana he shipped a cougar, five black bears, seven red wolves, and two rare ivory-billed woodpeckers. In 1906 Lilly decided to seek greener pastures and left Louisiana for the Big Thicket of east Texas. There he was successful in killing a number of bears, and his reputation spread. He drifted into Mexico and spent many years in Arizona and New Mexico conducting predator control for ranchers and the government.

Lilly's legendary status was due in part to his peculiar looks and habits. President Roosevelt wrote of him: "He has a wild, gentle face, with blue eyes and full beard; he is a religious fanatic and is as hardy as a bear or elk, literally caring nothing for fatigue and exposure which we couldn't stand at all. . . . often he would be on the trail of his quarry for days at a time, lying down to sleep wherever night overtook him."

Lilly would not raise a hand to work on Sunday. He never cursed, smoked, or drank alcohol or coffee. He was known to subsist for days in the wilderness with only a sack of corn meal. Ben preferred to sleep and eat outdoors even when amenities were available. Laden with bearskins and live cougar kittens, his brief and infrequent visits to towns only enhanced his enigmatic aura. Given the opportunity in a crowd, he was known to promote his own heroic folklore.

Lilly died in Grant County, New Mexico, in 1936, about eighty years old. His epitaph in the Old Silver City Cemetery reads, "Ben Lilly—Lover of the Great Outdoors." By modern standards, the inscription would contradict his lifestyle of the relentless pursuit of apex predators. He was, by any standard, cast of a different metal.

Monuments to Ben Lilly have been erected in Mer Rouge and in New Mexico's Gila National Forest. Recently, the Ben Lilly Conservation Area was established in Morehouse Parish along Bayou Bartholomew.

[A version of this story first appeared in *KnowLouisiana Encyclopedia,* 2013.]

4
REFLECTIONS, RUMINATIONS, AND TRIBULATIONS

Bayou Memories II

Most people carry childhood memories throughout their lives. Some are heavy and burdensome like a big sack of pears from the old tree on the homeplace, while others are light as owl feathers. Those that endure cannot be separated from and are always tied to a place in the mind's eye. In Louisiana the place is often a familiar bayou, river, or lake that lay on the landscape of our youth. Many of my earliest memories germinated on or near the ever-varying waters of Bayou D'Arbonne.

I distinctly remember when I was four years old going with my father in his early '50s Ford car down to the bayou bank at the mouth of Rocky Branch where he target practiced with his Marlin .30-30 deer rifle. It is not a pleasant memory. Much to the disgust of my father, the explosions terrified me, and I ran to cover my ears, crouched on the hump of the backseat floorboard. The irony of this occasion is expressed in the fact that I can no longer hear pine warblers after shooting firearms for six decades.

I remember, soon after the target-practice incident, sitting on an overturned lard bucket in a cypress john boat as my grandfather paddled down the bayou. We were supposed to be going fishing, but for the life of me I could not understand how we were going to catch fish without bait. When I mentioned this problem earlier, Papaw didn't seem to be worried. He paddled along leisurely until abruptly changing course to tie up to the overhanging limbs of a water elm. By then I was becoming anxious to fish, and when he reached down for the morning newspaper that was just too much for me. A patient, "Hold on, boy" was less reprimand than I deserved as his plan played out. He wrapped several sheets of the newspaper around a cane pole and tied it on with an old shoestring. Then he struck a kitchen match by rubbing it along the side of his boot and set the newspaper afire. Abruptly standing in the low-sided bateau, he raised the torch overhead and began poking a dinner-plate-sized red-wasp nest that I had until this point not been aware of. The idea was to kill the adult wasps with flames and use their larvae for fish bait. For a prescient five-year-old boy the incident was thought provoking to say the least as I received an early lessen in bayou improvisation. I don't remember if we caught supper.

In a few more years I was let loose on the bayou and its swamp that beckoned from the bottom of the red-clay hill where we lived. While summer camping in the heart of the swamp with neighborhood boys, I learned from Old Tink how to fish for giant gar in the shallow pools by using a long, cottonwood pole with a short, wire leader and huge 20/0 hooks baited with whole gaspergou. The floating pole was tossed in the bayou and would stand on end when the bait was struck. Eating the hot, deep-fried chunks of gar on the bayou bank at sunset branded memories on the sunburned hides of hungry boys. So did the haunting calls of barred owls and the screams of panthers that always turned out to be my great-uncle Clarence, who would hike barefoot through the swamp to our remote campsite and arrive just after we collapsed exhausted on army cots draped with mosquito bars.

If there is a gene that controls the propensity to enjoy fishing and hunting, I inherited the dominant allele for that trait. My first memories of deer hunting included a black-and-white paint stallion and a pack of gangly walker dogs with names like "Popeye" and "Queenie." They belonged to Uncle Tonce, who coordinated our hunts in the swamp where deer populations had finally recovered with the aid of restocking after decades of absence because of unregulated and illegal hunting. A plan of action was concocted each morning long before daylight around a breakfast table laden with biscuits, mayhaw jelly, bacon, and fried yard eggs. Standers for the day were advised of their assignments for the hunt at places where deer were known to travel. Boys were instructed which of the hounds to catch and restrain with grass-rope leashes from the howling cacophony of the dog pens. The standers left walking to their positions in the swamp with flickering flashlights and never enough clothes in the predawn darkness. At first light, Uncle Tonce saddled the restless, snorting stallion and loosed the frenzied hounds. His job was to lead them to a predetermined thicket where he hoped to jump a deer and then to follow them during the chase, heading them off when possible if the deer decided to swim the bayou. Meanwhile, the emotions of the hopeful standers varied with the volume and proximity of the crying pack. Are they coming this way, or did they backtrack across Dry Slough? Is the quarry just an old barren doe or the heavy antlered buck with drop tines that cousin Michael missed last year on the ammonia pipeline? These memories stuck, as did the scenario of grown men arguing over the proper way to hang a deer for skinning—head up or head down. Or of the image of my father walking down the highline toward me with snow on the ground and the legs of a deer hanging out each side of his hunting coat. Or of toting burlap sacks full of chopped sweet potatoes to Wolf Brake to illegally bait wood ducks that never came.

Many years later, when I became manager of the national wildlife refuge that now encompasses most of the D'Arbonne Swamp, it occurred to me that most of my early memories of the area involved consumptive uses—activities that consumed natural resources of the bayou and swamp. Whether trapping raccoons for high-school spending money, setting trotlines for catfish in spring backwater, or gathering muscadines for autumn jelly, the bounty was underappreciated at the time. Though my activities as an individual were

sustainable then, times have changed. Now with the paint still drying on my bigger, 3-D picture, it is early memories that will ensure
advocacy and gratitude for the treasures that remain.

Eclipse and Science

What a glorious event it was, this eclipse of 2017. With family and
friends, we gathered in a remote sagebrush meadow along the South
Fork of the Payette River in Idaho, a stream fresh from the Saw-
tooth Wilderness that still carried snowmelt even on this late summer morning. Our chosen viewing post fell directly in the path of
the umbra—the shadow projected when the sun is entirely blocked
by the moon. In the United States this roughly seventy-mile-wide
path arcing from Salem, Oregon, to Charleston, South Carolina,
was the only place to see the total eclipse of August 21. Other areas
experienced varying degrees of a partial eclipse.

Our group was prepared with certified eclipse-watching glasses.
Lawn chairs arranged in a semicircle were full of giggling children
who didn't quite understand the significance of the occasion. Suddenly, an adult staring at the sun said, "It's begun!" Indeed it had.
From this moment until it was over everyone spoke in hushed
tones—instinctively and without prompting, even the kids. As the
moon consumed the sun, the bright blue sky dimmed along with
birdsong until stars appeared that were linked by the loops of awak-
ened bats. The air chilled quickly. And then the unearthly totality
occurred—for two minutes and thirteen seconds. The safety glasses
could come off. The dark orb wore a glistening, platinum ring of
coronal streamers. One could not look away. Too soon, a spear of
sunlight flashed from the right side of the sphere and the associated
phenomena reversed direction.

For several days afterwards I was vexed by something pertaining
to this incident that I couldn't quite extract from my subconscious
meanders. It seemed to relate to current affairs in this country. Finally the boredom of driving through a three-hundred-mile-long

Nebraska cornfield shook it loose: Millions of people in the United State eagerly altered their lives to participate in the eclipse. Cumulatively, they drove and flew millions of miles to be beneath the path of the umbra. They spent incalculable amounts of money on travel, hotels, and food. It was major network news. All of this activity was planned around precise published data about the eclipse. At our location we knew that the event would begin at 11:28 a.m. Mountain Time and that it would last two minutes and thirteen seconds. Starting time and duration varied across the country, but the exact figures at each place were precisely known. Who figures this stuff out? The answer, of course, is scientists.

The conundrum is that a significant percentage of people in the United States don't believe in objective science if it contradicts their subjective worldview. No better example exists than in the widespread denial of human-caused climate change and its accompanying anti-science rhetoric. I wonder how many of these deniers went out of their way to view the eclipse that was known to them in such detail only by the exquisite calculations and forecasts of the same scientists they otherwise denigrate. It is as if the dark eclipse blinders are never taken off.

Alligator Allure

As apex predators, carnivorous species that lurk at the top of the food web, alligators toy with our psyche not unlike the mythological Sirens. Once exposed, we are vulnerable. Alligators, as we know them, have existed for sixty-five million years or so. As a species, we humans have shared common ecosystems with American alligators for no more than thirteen thousand years, and most of what we learned of them in that time has been forgotten. Yet within the brief era of recorded history, humans have accumulated a large body of information pertaining to alligators. As a method of assessing its significance and the permeation of alligators into all aspects of our modern culture, consider that a recent Google search of the word

Alligator Allure

"alligator" yields 57,600,000 results in 0.18 seconds. How is this to be interpreted? Which of the results are germane? Is it the article that describes how President Hoover's son, Allan, allowed his two pet alligators to wander around inside the White House? Does the journal report of a medical discovery that alligator blood products may successfully treat diabetics pass muster? Is it the website for the Alligator Warrior Festival in Lake City, Florida? Or the scientific paper that assesses the impacts of sea-level rise on coastal alligators? Assuredly, only those search results that expose a connection between alligators and people should be considered important. That would leave all 57,600,000 as a measure of their standing. And so the fascination continues.

[Adapted from my *American Alligator: Ancient Predator in the Modern World* (University Press of Florida, 2013)].

Ode Trip

People who don't live in East Carroll Parish, Louisiana, go there for various reasons. Most are likely passing through on Highway 65 heading north or south for destinations far from the rural, agrarian landscape. It's been that way for a while. Certainly during the Civil War, General Grant had no intention of lingering when he set up shop there and impressed the local slaves to dig a canal through Sharkey clay soils in order to bypass Vicksburg and the entrenched Confederate Army, who were aggravating the dickens out of Yankees trying to float past their fortress above the Mississippi River. But I digress. Of the modern visitors, few show up on motorcycles, especially one made in Germany. However, I expect that once in a while it happens. Surely, though, no one has ever gone to East Carroll Parish on a German motorcycle for the reason I did recently. I was conducting a census for the University of Texas. It had nothing to do with the demographics of the poverty-stricken parish or the forecast of cotton and corn yields after the wet spring. I was merely documenting the presence or absence of the likes of widow skimmers, swamp darners, citrine forktails, and wandering gliders. Of course these names belong to dragonflies and damselflies (taxonomic Class Odonata, aka Odes). It seems that of the sixty-four Louisiana parishes none have fewer documented species of these insects than East Carroll—eleven to be exact. This compares to sixty-four species found in East Feliciana Parish. Why so few species in East Carroll? Probably because not many people have searched for them. The lack of habitat diversity in the chemical-laden sea of agriculture may also be a factor. I am sorry to say that my survey did not increase the known species list, but I am encouraged and will try again. Lest you think this is some esoteric scientific study funded by taxpayers, be assured that nothing is farther from the truth. It depends on volunteer citizen scientists. Oh, as for the German motorcycle—it was my conveyance of the day only because it gets sixty-five miles per gallon.

Bayou Mud

In the beginning there was mud, bayou mud, molding the ontogeny of the boy. The summer of 1964 was the beginning for me, when I was thirteen and loosed upon the bayou and its swamp for the first time. Rules were minimal. Even the "be home by dark" decree was rescinded by autumn. An intimacy with local geology started with the half-mile barefoot walk from home to the bayou. Descending toward the stream, old sandy-clay soils of the Tertiary epoch gave way to new, dark silt-loams of the active floodplain. Alluvial mud comprised of this younger dirt and swamp water was adjacent to and under the bayou, wherever it happened to be in its seasonal wanderings. Extrasensory when encountered, mud became the pilings upon which the boy constructed his definition of this place.

The odor of summer mud imprints first. It is pungent with a bit of putrid decay, the tell-tale exhaust of the solar-powered, biological engine that runs the swamp. In time the odor became alluring, even haunting. Eventually, it settled onto my first cranial nerve and became a memory aroma. From this point forward for years to come, the odor overwhelmed the man the boy had become with distinct images of adolescent experiences panned across the screen of his mind. They appear most vividly when I approach the water's edge after having been away from Louisiana bayous for a long time—seining a sandbar on a stormy night, checking trotlines at dawn with great anticipation.

For the man, the tactile experiences of mud fill the bookshelves of his memory only second in volume to odors. Imagine the rough, sunbaked texture of cracked mud in mid-August. Fissures in gray, geometric patterns irritate even the late summer soles of his bare feet. Imagine also on this day, when the humidity is two-thirds a temperature that approaches the century mark, a sensation of cold, silky velvet that envelopes the lower extremities as deep as the boy can bury them by wading out and wiggling his toes mole-like toward the center of the earth.

Bayou mud yet defines his place.

Deer Hunting

Recently I walked across the street from my house and killed a deer on the edge of the D'Arbonne Swamp. The land there is in the process of producing its third forest in historic times. One hundred years ago my great-grandfather tried to feed his large family by growing corn and cotton on the marginal soils of this Pleistocene terrace. Eventually they pretty much starved out, and it was root hog or die for those in my grandfather's generation. One thing is certain. Neither of these close relatives could have supplemented their larder with venison backstraps from the property as I did. In their day deer had been eliminated from this area and indeed from much of the eastern United States. The species that had once played the same vital role for eastern Native American cultures as bison did for those in the west survived only in the most remote and inaccessible regions. Habitat modification and large-scale, unregulated hunting initiated a population plunge that ended only in the early twentieth century.

The onset of scientific wildlife management, including applied research and restrictive harvest regulations, has resulted in a dramatic recovery of the species throughout its range. At an estimated population of 30 million, more white-tailed deer live in North America now than at any time in the past. Last year in Union Parish where I shot the deer, hunters according to Louisiana Department of Wildlife and Fisheries records harvested 5,630 deer, or 1 deer per 85 forested acres. The total harvest for all Louisiana was 153,500. For hunters the resurgence of deer has been a welcomed phenomenon.

The news though is not all good these days. Deer populations in many areas of the country are now unnaturally high and unsustainable. Deer predators no longer exist in numbers sufficient to keep herds in balance with habitat. Hunting as a management tool to keep populations in check is declining. Managers are concerned that widespread habitat destruction and die-offs may occur in some regions. This scenario is not likely to play out in Louisiana any time soon because of staunch hunting traditions. If it does at some point, the irony will be that venison backstraps will once again become scarce in Union Parish because of a lack of hunting instead of too much of that activity.

Bad Ideas

When it comes to the natural world, we don't know what we don't know. Trouble jumps up about the time we think we've got it all figured out. There are plenty of examples of well-intentioned human actions that have caused environmental chaos. One pertains to recent attitudes concerning wildfires. For a century, fires on natural landscapes were thought to be unmitigated disasters. Tremendous efforts went into fire prevention and suppression across the country. Smoky Bear taught generations of children that fire is bad. This ill-informed position, by failing to recognize that fire is a natural part of many ecosystems, has led to very unnatural conditions in many regions. The consequences are that some plants, with animals that depend on them, have almost disappeared because they can't live without occasional fire in their habitat. Some seeds don't germinate unless released by heat. Prairies turn to shrubby thickets if not kept in check by fire. When fires do occur in areas after long periods of fire suppression, they often are so hot as to cause serious environmental damage.

Another example involves predator control. For many years governments had formal programs to eradicate predators that were thought to compete with human interests. By shooting and with the aid of poisons, large predators were totally eliminated from much of the country. Only recently and with continuing controversy has the healthy roll of predators in an ecosystem been recognized even by professionals in the field. These predators can range from gray wolves in the American West to alligator gar in Louisiana bayous.

The list of other bad ideas is long and includes the intentional introduction of invasive species such as kudzu and Chinese tallow without thinking of the consequences. Likewise, levees along the Mississippi River in south Louisiana were built for flood protection without ever considering that they would contribute to the loss of our state's critical wetlands. Until we recognize as a society that we don't know what we don't know, and that good science is the path to new knowledge, we will continue to be surprised by our blunders.

Tragic Targets

Recently a local man with too much time on his hands chose to spend a beautiful spring morning driving the back roads of Franklin Parish armed with a .22 caliber rifle. At some point he spotted a spectacular adult bald eagle perched in a tree and shot it. He cut off the bird's magnificent white head and threw the carcass in a ditch.

As a boy growing up in the outdoors of north Louisiana I never once got a glimpse of our national bird in the wild. There were almost none to see. Continental populations of bald eagles had been decimated by the pesticide DDT; when leaching into the food web,

Bald Eagle

it resulted in eggshells too thin to support the embryos inside. In Louisiana the few surviving birds nested in remote swamps along the coast. DDT was eventually banned, and bald eagles were placed under the protective umbrella of the Endangered Species Act. The slow recovery commenced. In the early 1970s I began conducting wintering eagle surveys for the U.S. Fish and Wildlife Service. Each year a growing number of bald eagles from the North overwintered in areas of prime habitat such as the region that would become the Mollicy Unit of Upper Ouachita National Wildlife Refuge and the Boeuf River bottoms. Still, a successful nest in north Louisiana was years in the future. Since that time, the recovery of bald-eagle populations across America has become one of our nation's most rewarding conservation success stories, and eagle nests are now fairly common in our area.

The man who killed the Franklin Parish eagle was apprehended and sentenced in federal court. He received a substantial fine and loss of privileges. A harsh penalty reflects the seriousness of the crime and intends to serve as a deterrent to depraved behavior. The message is not getting out. Soon afterwards a bald eagle was found walking along a road in Union Parish. He was captured and an examination yielded a wing broken from gunshot. Amputation was necessary, and the wild bird will spend the rest of his life in a cage. What does this say about our culture that we would foster such conduct with less than outrage?

Balloon Releases

Louisiana. Across the country we are known for our moss-draped cypress trees, slow meandering bayous, antebellum homes, spicy cooking—and litter. In spite of our wonderful assets, our litter liability is one thing most visitors remember. Everyone is affected. All other things being equal and given a choice, would you rather live and work in an area that is trashed out, or in one that isn't? It's one of those ephemeral quality-of-life issues.

There is a form of littering that many people innocently condone without thinking. I find it in the most remote areas of our state, in the middle of roadless floodplain forests and far out in the marshes, in areas where other litter is uncommon. In these settings it is especially garish and obtrusive. I am referring to balloons, the helium-filled ones that are released at promotional and political events, ball games, and parties. When released balloons are out of sight, they should not be out of mind—what goes up must come down. Millions are released each year. Latex balloons, which are made from the sap of rubber trees, are usually biodegradable and cause less of an impact than foil or Mylar balloons, which persist in the environment. Attachments such as ribbons and string are particular problems as they sometimes entangle birds and other wildlife. In coastal areas, sea turtles, dolphins, whales, and fish ingest balloons, which they may mistake for jellyfish, a favorite food.

So how can we clean up this small but important piece of our litter problem? Obviously the first step is not to release balloons outdoors. We can encourage companies, schools, political and community organizations to use an alternative form of advertising or celebration. Above all, we can educate, for a littered state is an uneducated state.

Environmental Controversy

The existing state of affairs regarding America's environmental issues mirrors that of most high-profile topics on the political agenda. As with health care, taxation, and foreign policy, there is no longer a middle-ground constituency that can be heard above the imbroglio of camps so disparate in their opinions as to seem to be from different planets. Indeed, an alien could quickly get up to speed on the status quo by reading the comments posted under online news articles about current environmental issues. The loudest opinions are beyond fractious as posters oppugn opposing views with vitriol instead of rationale. On the national scene, climate-change, energy-policy, and endangered-species agendas are ground up by both conservatives and progressives like a road-kill armadillo on I-49. In

Louisiana we personalize the list to include deepwater drilling, natural-gas fracking, and abuse of our aquifers. Even an environmental calamity such as the BP oil spill failed to unite factions for the common good in the long term. It defies common sense. Why should political, religious, social, or occupational affiliation matter when it comes to advocacy for healthy life-support systems? An applicable quote from the late Senator Patrick Moynihan reads, "Everyone is entitled to his own opinion, but not his own facts." As for the venomous atmosphere surrounding the debates, Joseph Addison, the early eighteenth-century English essayist, wrote: "A man must be both stupid and uncharitable who believes there is no virtue or truth but on his own side."

Grandson and Population Growth

How is it that the presence of a gloriously innocent two-year-old grandson can make the environmental predictions of the future seem a bit more relevant? What's in store for him? As I began writing this story, the world-population clock was approaching 7.5 billion. In the United States the net gain is one new person every twelve seconds. Although the rate of population growth on this planet is declining, it is not predicted to stabilize until after the year 2200, at over 10 billion persons. This could be within the lifetime of my grandson's grandson. All of these people, kinfolks included, will seek the material things that we can no longer seem to live without, not to mention an ample supply of food, water, and shelter.

Within the realm of environmental degradation, gloomy news is so commonplace that we tend to become desensitized and point our antennae toward more pleasant vibrations. Who wants to hear that 6.5 million acres of forests have already been lost this year? It is not pleasant to think that 70 million barrels of oil will be pumped from the ground today and rapidly consumed, or that 5.3 million tons of toxic chemicals have been released into the atmosphere so far this year. After all, what can we as individuals do about it? The answer of course is that cooperatively we can do a lot if we set our minds to it. First though we must realize that we and our descendants are not somehow separate from the natural world but totally dependent on it. Once this basic concept of an environmental conscience sinks in, we will work through the approaching population bottleneck. Collectively, we are a long way from developing this mind-set. As I watch this precious child, my genetic link to the future, chase lizards on the long front porch, I can't help but worry about what will be left for him and his grandchildren if change doesn't begin soon.

Reverse Invasive Species

One of the most challenging issues on American conservation agendas involves the control of invasive species, those plants and animals native to other parts of the world that disrupt ecosystems when introduced accidentally or in some cases intentionally. Invasive species impact natural energy cycles, wreak havoc with native species, and have economic impacts measured in billions of dollars annually. In Louisiana we confront water hyacinth and giant salvinia that clog waterways, Chinese tallow trees that preempt forest habitat, Formosan termites that eat live trees, and nutria that consume marshlands. The natural homes of these pests are thousands of miles from the Bayou State.

An aspect of the invasive-species issue that is rarely mentioned in the United States pertains to those species that originate here but become problems elsewhere. In other words, not only are we recipients of invasive species, we are also donors to other parts of the world. A classic example is North American gray squirrels that were released in the British Isles. There they have few natural predators; they out-compete the native red squirrels, and harbor a virus fatal to the reds. Eradication efforts to date have been unsuccessful, and British celebrity chefs are now promoting the idea of eating gray squirrels similar to the efforts of Cajun chefs to control nutria.

Additionally, mink, Canada goose, ruddy duck, rainbow trout, several moth species, jewelweed, evening primrose, and duckweed from America are considered invasive species in the British Isles.

American bullfrogs and mosquitofish (a small minnow common in Louisiana bayous) cause problems in Asia. Broomsedge grass from America displaces native plants in Japan and Australia. The signal crawfish, a species found west of the Rocky Mountains, is in the process of ousting native crawfish across Europe.

The United States has even become a secondary source of invasive species. Fire ants arrived here from South America in the 1930s and now have a six-billion-dollar annual economic impact. Genetic tests on fire ants in the recently invaded countries of China and Australia indicate they came from the United States rather than South America. Whether we are recipients or donors of invasive species, most problems result from human tinkering with nature, good intentions notwithstanding.

Coastal Wetlands

"It's going fast!" shout the headlines, but it's not all gone yet, this vast and wonderfully productive fringe of wetlands that clings precariously to Louisiana's coastline. Geologically young, they have been battered since infancy by a giant feral river with a mouth more like the severed tail of a lizard that whipsawed back and forth, leaving scattered deltas in its wake. This mother river was stern, seemingly whimsical in her behavior, and yielded only to the force of gravity. A fixed point would dress as a brackish marsh one decade, main stem of the prowling river the next. She was nothing, though, if not nurturing. Almost innocuously she slipped from her banks in spring floods to gather the richest nutrients North America could offer and rushed them downstream to fatten the wetlands during annual silt-laden orgies of gluttony. In this manner the coastal wetlands flourished, and not even the likes of a Katrina hurricane upset the dynamics for long. As natural ingredients in the overall system, powerful storms were nothing more than a brisk stirring of the environmental gumbo.

For thousands of years B.C., Louisiana's coastal wetlands were broadly static, meaning that the functions and processes operated to sustain the system. The sheer volume of biomass generated then is difficult to imagine. For instance, a recipe to cook up the historical wetlands ecosystem would include most of the country's shrimp and other seafood, and the vast majority of the continent's millions of wintering waterfowl. The decline began with the end of the B.C. days, "B.C." meaning "Before the Corps of Engineers." Our cries to subjugate the haughty manners of the Mississippi River resulted in a bureaucracy that did just that—at least in the short term. Dams, levees, and continuously dredged ditch-like channels barred nutrient access to wetlands and dumped them into the deep-sea abyss beyond the continental shelf. The wetlands began to starve and slumped into a posture of unnaturally rapid subsidence. Poison in the form of saltwater where it should not be was introduced into the system with the development of the oil and gas industry. Thousands of miles of pipelines and access canals provided a conduit for seawater to flow landward and effectively sterilize fresh and brackish marshes of vegetation that held the world together. Bayous that served as low-gradient capillaries were overwhelmed by the saline incursions. Biomass and biodiversity plummeted.

Even without the proven values of coastal wetlands to our own shortsighted species, their intrinsic aesthetic worth merits a call for action. We are smart enough to help nature's remarkable resiliency overcome the present "almost gone" status, even in spite of the BP oil-spill disaster. It will, however, require a collective commitment very soon from all who read the headlines. Feeling like the first cool dry breeze of autumn after a long hot summer, a touch of momentum is in the air.

Habitat Loss

Louisiana license plates proclaim our state to be the "Sportsman's Paradise." The inference is that Louisiana harbors a cornucopia of fish and wildlife resources. Compared to many parts of the country, it still does. We are yet blessed with an abundance of wildlife. That is not to say that all is stable on the home front. All plants and ani-

Magnolia Seedpod

mals from cow oaks to bullfrogs are dependent on specific habitats within broader ecosystems. Degraded or destroyed habitats always result in a decline or disappearance of those species that naturally live there. Most large-scale disturbance to habitats is insidious because it happens slowly over many years. Clearing of the vast bottomland hardwood forests of the Mississippi River floodplain has been ongoing since before the Civil War, and though there were pulses of peak activity, most recently in the 1970s, the existing forests of this type are survivors of erratic patterns of destruction over 150 years.

By far the highest-profile issue of habitat loss in the state involves the disappearance of coastal wetlands. Bumpers stickers seen around the country cry out to "Save America's Wetlands." Only the oldest Louisiana citizens alive today who experienced that ecosystem can fully appreciate the extent of the losses. For the rest of us, the sheer biomass of life, from shrimp to waterfowl, that once thrived there is difficult to comprehend. With few exceptions, the loss of coastal wetland habitat occurred in the last seventy-five years primarily as a result oil and gas extraction activities, and flood control and navigation projects.

The loss of bottomland hardwood forests and coastal wetlands each reached a critical point that triggered a public outcry to stop the habitat loss and remediate the widespread damage. Much of the remaining bottomland hardwood forests was preserved in national wildlife refuges and state wildlife management areas. Federal agricultural programs provided incentives for landowners to reforest unproductive farmlands. On the coast, massive planning efforts and hundreds of millions of dollars are involved in the effort to stem the loss of wetlands.

Another major ecosystem in Louisiana has been drastically altered in the last fifty years but not yet to the point of widespread citizen concern. Upland forests in the state once comprised a diverse, vibrant ecosystem scattered over millions of acres. Where once lived chinquapins, lady's-slipper orchids, and box turtles, rows of genetically altered, monotypic pine trees now cover the landscape from horizon to horizon. Perhaps there has not been an outcry to protect a part of this unique habitat because we have bought the company line that these farms are actually forests, which in fact they only remotely resemble. It is a literal case of not being able to see a forest for the trees. Unless we begin to value all of the various natural areas in the state for their intrinsic worth, our conservation efforts will be as outdated as an old license-plate slogan.

Little Missouri

I have boyhood memories of motoring with my parents along a stretch of arrow-straight, asphalt highway as it passed through a vast and seemingly desolate swamp in north Louisiana. Understory palmetto fronds lent a tropical ambience and obscured the ground under the tall, dark trees. The road was an incision in the forested canopy.

At the time it was one of the largest remaining tracts of bottomland hardwood forest in the region. Roughly bordered by the Arkansas state line on the north, the Richland Parish line on the south, Bayou Bonne Idee on the west, and Boeuf River on the east, the swamp is bisected by Louisiana Highway 2 between Mer Rouge and Oak Grove. When I first passed through the area in the early 1960s, it was known as the Bonne Idee Swamp or upper Boeuf River Swamp, and its days were numbered.

As is common in most intact swamps, the biodiversity and abundance of natural resources were great. Native American artifacts sprinkled over the ridges indicate that humans found nurture there for millennia. Well over a century ago, immigrants of European origin settled around the edges of the forest and tapped its wealth mainly by hunting. One of the country's most legendary hunters, Ben Lilly, began his career here at the end of the nineteenth century in pursuit of bears that flourished in the dense canebrakes. The reputation of the area as prime hunting ground grew to the extent that in a few years chartered trains were bringing hunting parties from Chicago for weeks-long adventures.

The swamp survived in a sea of agriculture as long as it did because it was just that—a swamp, low in elevation and frequently flooded. But beginning in the mid-1960s new pioneers came to the region, bringing the latest farming technology and ample determination. They commenced to clear and drain the entire swamp. Soybeans and later rice replaced willow oaks and cypress brakes. Hundreds of thousands of forest-dwelling songbirds went elsewhere or nowhere. Bears could not find refuge in the likes of grain silos. So many of the new farmers came from southeast Missouri that the area became known as "Little Missouri."

Today, from a conservation perspective, the situation is improving a bit in Little Missouri. Landowners are beginning to realize the recreational value of natural areas. With government incentives, several thousand acres of marginal farmlands have been reforested. If they endure, the songbirds will return, and already wandering bears are monitoring the changes.

Oiled Birds

The ongoing tragedy resulting from the Deepwater Horizon oil spill off the Louisiana coast has generated an unprecedented wildlife-oriented media blitz. Of all the images that zip around the world and spill from TV sets to drench our hearts, none are more emblematic than oil-soaked pelicans, wings spread and mouth agape in obvious agony. For most of us these photos evoke emotions of pity first, closely followed by anger for the perpetrators of such suffering on innocent creatures. No wonder then that the so-called rescue groups that clean and release oiled birds are the white knights in the eye of the media and most of the general public. It should be so if for no other reason than most of the people involved are volunteers.

There is another angle of the story though that rarely makes page one, and it involves the actual effectiveness of rehab activities. Many wildlife professionals believe that the impacts are negligible. The few relevant studies indicate that the survival rate of cleaned and released birds is extremely low in most cases. Consider the stress involved in capture, transportation, detergent baths, and recurring handling by creatures perceived as deadly predators—that is, humans.

Additionally, the cost of rehab even when conducted by volunteers can be as high as thousands of dollars per released bird. Wouldn't it be better just to euthanize the poor creatures? Probably, except in the case of endangered species when every individual is important to the survival of the species. Certainly, if the costs of bird rescue activities could be channeled to long-term habitat protection instead. There is no doubt that a million dollars spent on the purchase and perpetual safeguarding of a thousand acres of marshlands would accrue vastly more benefits for wildlife of all types than if the same million dollars were spent on cleaning and releasing oiled birds.

It will not happen, though. The public will not allow oiled birds to be systematically killed, even humanely, regardless of the real efficacy of rescue efforts. An even greater irony exists in the positive public-relations aspect of bird-rescue activities being enjoyed by the company that caused the spill, BP in this case. To really make things right, shouldn't we demand that they purchase for us that thousand acres of marshland several times over in addition to cleaning birds?

Wilderness

As a species we humans are infamous for behavior not conducive to our own long-term well-being. Consider the frequency of wars, the unbridled depletion of earth's finite resources, and the "me now" attitude of our consumptive society. There are, however, shining examples of farsightedness in America, even in the halls of Congress. A prime example is the Wilderness Act of 1964.

This law created a way to designate and protect a system of undeveloped lands called "wilderness areas" across the country. It states the places should be "an area where the earth and its community of life are untrammeled by man, where man himself is a visitor who does not remain." It is America's highest form of land protection as roads, vehicles, and permanent structures are prohibited, as are commercial activities such as mining and logging. Such restrictions should not convey the idea that the areas are unused. Millions of

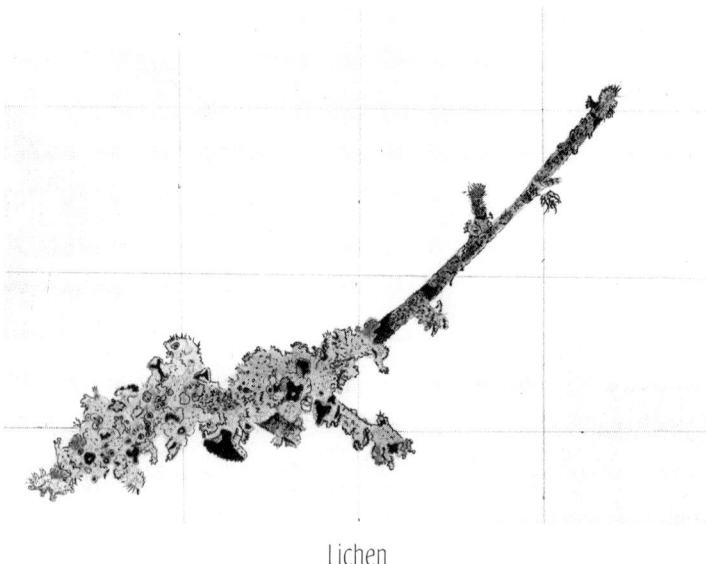

Lichen

annual visitors hike, camp, hunt, and fish on the lands and waters. Today, more than 750 official wilderness areas are found in forty-four states and encompass almost 110 million acres.

Detractors claim that too much of the country is tied up in wilderness areas that shackle their ability to exploit natural resources. In fact, just over 4 percent of the United States is designated wilderness. Cynics ignore very real ecosystem services afforded by wilderness areas such as the protection of watersheds that provide clean drinking water, the filtration of air we breathe, and the protection of wildlife habitat. From an economic perspective, wilderness areas boost local economies with tourism and recreation dollars. For many of us the aesthetic worth of visiting a wilderness area, or just knowing they exist in an otherwise chaotic world, is incalculable.

In Louisiana, most people are unaware of our three official wilderness areas. Congress designated the Breton Wilderness in 1975, comprising 5,000 acres of the Breton and Chandeleur islands and surrounding waters just east of the mouth of the Mississippi River. Critical habitat for thousands of seabirds, they are also frequented by birdwatchers and surf fishermen. Kisatchie Hills Wilderness (1980) encompasses 8,701 acres managed by the U.S. Forest Service in the central part of the state. Historic longleaf pine forests with their associated flora and fauna are protected and threaded with popular hiking trails. My favorite is the Lacassine Wilderness (1976), 3,346 acres of pristine freshwater marsh in the Lacassine National Wildlife Refuge of southwest Louisiana. I had the privilege of overseeing the area many years ago and to witness the lushness and diversity of nature when untrammeled. It is a relic and example of what was once common. Waterfowl hunters and fishermen still relish their visits.

The Wilderness Act is not invincible. It was created by Congress and can thus be abolished or rendered impotent by that same body. Indeed, like a host of species that depend on it for survival, it has never been more endangered than now. The vision of our current policymakers seems to be dimming.

Dog Day

In the dog days of summer after the fresh-split firewood reeking with the sweet acerbity of tannin is stacked in a neat pile close by the house, we become crepuscular. Like certain amphibians striving to maintain a proper balance of body fluid and temperature, we venture forth into the out-of-doors only in the twilight hours of dawn and dusk, leaving behind our artificial cocoons of refrigerated and dehumidified air. Even the cicadas are now out of sorts, droning about their business at midday when a pregnant cloud passes in front of the sun. In the first slow light of morning we sip strong coffee on the back porch facing east toward the hardwood forest where the birdsong rises. The cardinal calls first; then the liquid flute of the wood thrush sounds from the understory. Thoreau wrote of the wood-thrush song, "Whenever a man hears it he is young, and Nature is in her spring; wherever he hears it, it is a new world and a free country, and the gates of Heaven are not shut against him." This cousin of the bluebird is now tracking the declining hours of daylight with a mysterious sundial embedded deep within his brain. On a night in mid-August he will flush at a silent alarm and begin a nocturnal journey that will end for the season in the coastal lowlands of Central America. As for the cardinal, he suffers not from innate wanderlust and with his kind will still be around to serve as Christmas ornaments in the vanishing dogwood trees of Union Parish. With coffee cups almost empty, we are surprised this morning by the running-late possum that peeks over the edge of the porch on his routine check of the bird feeders. We all conclude that, in spite of the bidding thrush and Thoreau's doggerel to the contrary, it is time to seek shelter again until the evening respite.

MBC Fund and LWCF

Louisiana is blessed with a number of national wildlife refuges that exist to conserve fish and wildlife and their habitat in perpetuity, or at least until the law changes. Most of these lands and waters are bought outright from willing sellers. Condemnation of private property no longer occurs in these matters. Especially in tight fiscal times, people may wonder if buying these properties is the best use of their federal tax dollars. Not to worry, as most of the acreage is not purchased with tax revenues but with two pots of money established by forward-thinking lawmakers.

The first involves the Migratory Bird Hunting and Conservation Stamp, better known as the duck stamp. Ninety-eight percent of duck-stamp receipts are placed in a pot called the Migratory Bird Conservation Fund. A commission oversees the distribution of the funds, which can only be spent for the purchase of wetlands and other migratory bird habitat. Since the duck-stamp act was passed in 1934, sales have generated over $850 million that have protected more than six million acres nationwide. In the fall of 2014, about $8 million of this money was used to add almost six thousand acres to the Mollicy Unit of Upper Ouachita National Wildlife Refuge.

A second source of funding is the Land and Water Conservation Fund created by Congress in 1965 to buy lands not only for federal refuges but also for national parks and national forests, along with other federal conservation needs. The money comes from leases that energy companies pay for oil and gas drilling rights on the Outer Continental Shelf. The idea was to use revenues from the depletion of one natural resource—offshore oil and gas—to support the conservation of other precious resources—our land and water. Unfortunately, this fund is much more politicized than the duck-stamp pot, and much of the $900 million that is generated annually is siphoned off for pet projects. In 2014 money from this account purchased a critical eight-hundred-acre addition to Black Bayou Lake National Wildlife Refuge.

There is never enough money in these pots to meet the demand across the nation, and in Louisiana we are very fortunate to receive a share. Unfortunately, there are always legislative threats to dismantle these essential conservation tools. A simple way to support them is to buy a duck stamp each year, even if you don't hunt.

Scorched-Earth Homebuilding

The defining elements of some our most beautiful, old neighborhoods are large, mature trees. In addition to being attractive, they moderate harsh temperatures and add monetary value to homes. Old photographs reveal that when most of the houses were built the yards did not harbor big trees, and the value and beauty of such specimens did not accrue until many years passed. Not much has changed in this arena in Louisiana.

For whatever reasons, developers and homebuilders in this region often adopt a scorched-earth policy when it comes to preparing house sites. Whether an entire subdivision or a single residence is the objective, the most popular plan of action seems to be to begin with bare soil devoid of anything green. Ironically, in many such cases beautiful trees are removed, trees similar to those that new homeowners will promptly try to establish at considerable expense in labor and cost, and even then the desired effects of mature specimens are fifty years in the future.

The scorched-earth practice is not standard procedure throughout the country and, like littering, is a self-inflicted injury especially common in our region. One doesn't have to travel far north, east, or west out of Louisiana to find examples of great efforts being made to preserve existing natural vegetation in areas of residential construction. And it does take effort. Simply wrapping orange flagging tape around the trunk of a tree is little more than symbolic. At a minimum, all activity beneath the drip line of existing trees must be prohibited. Occasionally, some trees are not bulldozed, but soil compaction around the trees during construction is a death sentence, the execution of which may take a few years as the tree slowly succumbs.

Those who promote economic development in our state often mention quality-of-life issues as critical factors in attracting new businesses. Notwithstanding the environmental benefits of responsible development, we must begin to think beyond littering in order to be successful in this arena.

Climate Change—Dying and Denying

In terms of ecological calamities, what does it take to get our attention? Does the air have to become so polluted that we must wear face masks? Do we have to completely run out of safe drinking water? What if every tree in Louisiana were to die stone-cold dead? On top of that, throw in half of all trees in Arkansas.

Similar things have already occurred in other areas. Air and water pollution occasionally make headlines, but few in Louisiana know of, much less appreciate, the tree mortality ongoing in several western states. To grasp the scale of the issue, consider that about half of Louisiana's land area is forested. That's 14 million acres. Arkansas has about 19 million acres of forestlands. If all of Louisiana's forests died and half of Arkansas's, the total would be about a million less than have already died—24.3 million acres, primarily in Colorado and Wyoming, where entire landscapes are shrouded in dead trees.

Those forests, mostly Ponderosa pine, spruce, and fir, are being devastated by a small beetle that for thousands of years has been a natural component of these ecosystems. The cause of their new explosive impacts is climate change. The average U.S. temperatures have increased almost two degrees in the last century, with most of the rise since 1970. This seemingly small upturn can have dramatic impacts in the biological world. Recent winters have not been cold enough to kill off the beetles, as happened previously as a form of natural control. They can now live farther north and at higher elevations; they can produce more generations of offspring in a year. And they are finding their host trees more vulnerable as they suffer from prolonged drought.

The result is a cascading apocalypse with effects far beyond the forests. Local economies are hard hit as tourism in ski resorts and national parks declines. I recently witnessed an example of the destruction when I visited one of my favorite Colorado campgrounds. Two years ago it was in a pristine setting with giant trees, a waterfall, and spectacular vistas. Today it is in the middle of a barren clear-cut as every tree died. People in the affected region may still argue the reason for climate change, but they are dead certain that it's a reality.

Permanence

While choosing a grave marker for my father, the matter of permanence surfaced. Enter then my acquired bewilderment and biases regarding the concept of what is permanent in the natural world. Permanence, says one authority, is a continuance in the same state or without any change that destroys the essential form or nature. For starters, is even a granite tombstone permanent?

As a rock, granite is indeed tough and durable. Dad's stone originated as a hot, rising mass of magma underneath northern Georgia, just east of General Sherman's infamous trail to the sea 325 million years later. At about nine miles below the surface it came to rest and cooled for a million years. A fusion of three minerals—white feldspar, gray glassy quartz, and black grains of mica—solidified into granite. The monolith continued to rise, and erosion wore away the overburden until the deposit was just beneath the earth's surface. Quarrying was then possible. Even though born of fire left over from the creation of the universe, granite according to the craftsmen can also be destroyed by heat.

People have been thinking about this issue of permanence for a while. The pre-Socratic philosopher Heraclitus opined that the only thing permanent is change. "Everything changes and nothing remains still," is Plato's interpretation of Heraclitus. Plato, though, leaned toward Parmenides' writings that denied the possibility of change and held that all things are permanent. Parmenides explained that appearances of change are only illusions and not true reality, sort of like the Hindu idea of the *atman* or soul as unchanging even as bodily forms are altered. Along that vein, Aristotle believed that all plants and animals, as well as humans, have an unchanging essence (that is, soul) in spite of transforming bodies. Some modern religions consider God's omnipresent love that permeates the world as the only thing permanent. Such liberating thoughts are pleasant to contemplate.

When considering the perspective of environmental permanence, it eases the abrasive noises in today's world to think in geologic scales until I remember that forthcoming eons do not encompass the life of my grandson and his future children. Even at my age it's difficult to contemplate farther.

Wise Ones

We are losing the old wise ones. Some of our most erudite naturalists never heard a professor's lecture or studied in a biology lab that reeked of formalin and moth balls. Still, they know the eddies where giant flathead catfish prowl and the ridgetop trails where coyotes forage. They possess the skill to weave hoop nets five feet in diameter to catch spawning buffalo in a spring freshet, and the knowledge to boil new coon traps in green walnut hulls to mask human odors. Most of these sages have a lifetime of experience in commercial fishing, trapping, or otherwise supporting themselves and their families by harvesting the natural products of our forests and streams. Those remaining were born before or during the Great Depression. Societal changes have eliminated the demand for their products, and in many cases so-called progress has liquidated the supply. The few who still practice their arts do so as an occasional sideline, reluctant to give up the feel of slow, heavy tugs on a trotline baited with gizzard shad and the pungent odor of a mink hide properly stretched. They remember the times when good luck afield meant making a land payment or another semester in college for a child, the first in the family to be so privileged. Keenly aware of natural phenomena because it affected their livelihood, they understand the consequences of a poor acorn crop and a late backwater, the implications of a dammed river and a clear-cut forest. More so than anyone, they know that we all are inextricably attached to the natural world and will sink or swim into the future depending on the depth of our commitment to its health. If this wisdom is lost, we will only acquire it again the hard way.

Worthless Plants and Animals

A famous conservationist, a man who is considered the father of modern wildlife management, understood the natural world better than most people of his time and indeed better than most people today almost seventy years after his death. Aldo Leopold understood that a species of plant or animal does not and cannot exist in isolation from other plants and animals. He knew that there are always critical connections between life forms that when severed tend to wreak havoc on the various plants and animals in their particular ecosystem. Leopold once said, "the last word in ignorance is the man who says of an animal or plant: 'What good is it?'" Unfortunately, many people still ask that selfish question in regard to species that are not warm and cuddly or have no apparent monetary value. For those, answers laden with aesthetic values or the merits

Fowler's Toad

of "keeping all the parts" are at right angles with their reality. These folks are only interested in conserving those species that bring them pleasure or profit. There are, however, plenty of examples of uncharismatic plants and animals that benefit humans. Horseshoe crabs would not even place in an invertebrate beauty contest, yet they are harvested by the millions for a blood component used to diagnose bacterial infections in humans. Many of our most important medicines are derived from otherwise unremarkable wild plants. Few people are aware that they or a loved one might soon be treated with drugs resulting from research on alligator blood. Those with diabetes or burn victims in particular stand to benefit from the reptiles. Each of these species depends on a web of life that includes many other plants and animals. We are aware of some of the connections between them, but surely there are many we have yet to comprehend. And, believe it or not, the well-being of humans is welded hard to that same well-being of many wild plants and animals, cuddly or not.

Bullets, Etc.

"I need some bullets, hurry up!" That was the garbled radio message I heard late one night in 1981 while lying in bed at the Lacassine National Wildlife Refuge. Be assured that my feet hit the floor of the assistant manager's house when those words soaked into my sleep-fogged mind. The call from refuge manager Brown was short, urgent, and indicated his location to be somewhere in the marsh near the intersection of Bayou Misere and the Intracoastal Waterway. We had been working shifts trying to catch poachers who shot deer from passing tugboats as they traveled through the refuge. This didn't sound good.

I ran across the headquarters compound, law-enforcement gear in tow (including plenty of bullets), and banged on the door of second assistant manager Hebert. In no time we had launched the Whaler from its cradle in the boathouse and were racing down the Mermentau River. Six miles later we came upon the scene of action.

You couldn't miss it. Brown had pulled over a large tugboat which was lit up like a small city, searchlights pointing in every direction. We managed to climb aboard and found the crew spread-eagled with Brown in the process of interrogation. He had heard shots just before the tug came down the bayou and assumed they were the culprits. The indignant crew insisted they were innocent and told of passing a small john boat just before being stopped. We searched every nook and cranny of the tug and came up empty-handed. There was nothing to do but turn them loose.

The investigation then turned toward the alleged john boat. If it existed, the boat had to be in the small, narrow bayou between us and its outlet in Lake Misere. A plan was made. Brown would make a twelve-mile run down the Intracoastal and enter the lake at the opposite end. Hebert and I would ease up the bayou and effectively bottle up our prey. We knew of a rickety houseboat anchored in a small cove off the bayou and thought that might be the base of operations for the john boat. As it turned out, it was. We gave Brown a head start since he had more water to plow. He left before it occurred to us to ask about the "out of bullets" message.

As we idled down the bayou, one of south Louisiana's renowned pea-soup fogs rolled in and cut visibility to a boat length. After about thirty minutes we neared the houseboat cove and could see the faint glow of lights and hear a generator. Someone was home. Shortly afterward, Brown's running lights appeared in the fog directly in front of us. We radioed him but got no answer. Suddenly, he raced forward straight toward the houseboat. Not having a clue what was going on, we followed him. Brown's boat struck the houseboat hard, propelling him between the split windshield and up on the back, covered porch of the houseboat. A man caught him just as he was about to continue into the water beyond. We later learned that Brown had seen two men skinning a deer on the back deck. His idea was to race forward, cut the engine, and jump aboard before they could throw the deer in. He waited a bit late to cut the engine and, had it not been for the outlaw catching him, he would have been ejected over the far railing.

In the meantime Hebert and I made a somewhat more controlled docking at the other end of the houseboat and stormed aboard. The front door was locked, but when I gave it a kick it was just like in the movies. The whole door fell inward on the floor. We stepped inside to a scene that would be expected at a chainsaw massacre. When the bad guys saw us coming they began slinging freshly dressed deer on the back porch into the houseboat. Blood was everywhere. Large smears on the floor led to carcasses under bunks and in corners. They were covered with sheets, towels, tablecloths, and underwear. After the initial shock, our first reaction was to secure the scene, that is, separate these knuckleheads from the numerous weapons lying about. The three poachers were soon tended, and almost as an afterthought I decided to take another look outside. A homemade "port-a-john" was perched on one corner of the front deck. As I walked by I noticed a man standing stiff in the shadows against the toilet. I instructed him (probably in less than a professional manner at this point) to come out with his hands up. He did not move. I wondered if he thought that I did not see him. I stepped closer and gave the order again. Same results. Finally, a prod with the end of a cold metallic object brought him forward. He seemed to be in some type of trance and remained in this condition for a while.

The final tally was four poachers, five freshly skinned deer, thirty-something rabbits packed in ice chests, and several guns, one with a recently filed serial number. The bad guys lost that time. The next morning as we were sorting out the contraband I remembered to ask Brown about the bullets. "Oh, when I grabbed my gear I just forgot to bring any last night," he said.

Hallowed Place

There is a hallowed place on the edge of the swamp where I walk in early morning. It slows my pace, wells up in my imagination, and forces me to ponder lives that are impossible to comprehend. Situated on the edge of a Pleistocene terrace, it is skirted by an old channel of Bayou D'Arbonne that has since moved a mile westward.

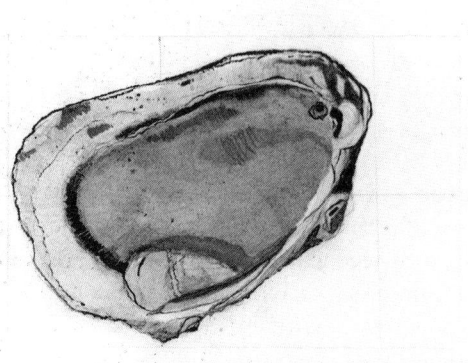

Mussel Shell

My great-grandfather called the site the "Big and Little Indian Camps." With his sons he cleared a few acres of the sandy ground and planted corn and cotton until the fertility played out. Since then the only crops have been a couple of generations of old field pines. It is now known that this remote spot was once a vibrant community whose occupants were members of a culture that we have labeled "Coles Creek." They lived their mysterious lives here in the D'Arbonne Swamp between 800 and 1,300 years ago. Of their traces, only broken potsherds, mussel shells, fire-cracked rocks, and small barbed dart points have been found.

When I stand among their ghosts, I try to imagine what in this environment they saw and heard. Certainly it was not the scar of a throbbing pipeline that now cleaves the camps. It was not the Japanese climbing fern nor the Chinese privet that has followed my invasive ancestors here. However, just down the hill there are large, old-growth trees—water oaks and sweetgums that are dying of old age. The Native Americans would have seen similar trees living out their natural lives. An uncommon plant that I am convinced was here then and that yet remains is yucca. Its rosettes of green daggers are sprinkled over the site. Yucca is often found on similar archaeological sites and is known to have been a source of materials for sandals and baskets. And it is possible, barely possible, that a few of the giant, lightning-crippled cypresses in the adjacent brake were young and vigorous when the village was present. The people surely heard the high-pitched, rattling call of the pileated woodpecker, as I do most mornings, and were familiar with its great ivory-billed cousin that I will never know except in old photographs. When I look up to envision the sky-darkening flocks of passenger pigeons that in their time mysteriously descended from autumn clouds in a roar of wings, I am distracted by the contrails of passenger jets. I suspect that, for a people who considered themselves a part of the landscape, the occasional howls of wolves and territorial screams of cougars were more of an aggravation than a source of fear. In that realm I am left with the euphonious music of opportunistic coyotes.

If this is a hallowed place, it broaches the matter of spirituality and religion. We have no way of knowing the religion of these Native Americans on the edge of the D'Arbonne Swamp or even if they had one. Based on what we have learned of their descendants after European contact several hundred years later, they likely were intensely religious with beliefs expressed in some form of animism or sun worship, practices roundly denigrated by prevailing Abrahamic religions. In some parallel universe where the Indian Camp inhabitants could view their home today to consider the damming of the bayou; the loss of passenger pigeons, wolves, and cougars to guns, traps, and poison; the demise of ivory-billed woodpeckers as a result of forest elimination, our definition of paganism would be a hard sell. Perhaps they would point to our sacred reflections in the bayou.

Arresting Poems

"Justice"

The ducks they killed are heaped in hidden piles
With limp wings askew like feathered pick-up sticks

They are not hungry these people
Bank president, successful realtor, government big-wig and the like

Still they purloin wildlife
As though their children were starving

As though environmental statutes are not real crimes
As though mallards will rise up from the marshes until the end of time

Illegal bait is concealed, escape routes are planned
Alibis are formulated just in case

With willful deceit they cover their trail like amateur bank robbers
And commend themselves on such cleverness behind locked gates

When it goes down, silence sweeps the thunder of shotgun volleys
Across the marsh and into the low-bellied clouds

The important people struggle to gain control of an unexpected imbroglio
In the blind the screed of a popinjay becomes belligerent

Egos swell and irrupt into a sad hilarity of threats
They will have my job; I will hear from their lawyers

When judgment comes, it is swift as a dozen teal
That strafe the bobbing decoys and vanish before a shot can be fired.

"Recreation"

Killing bald eagles, peregrine falcons, and whooping cranes is challenging
    and fun.
People have told me so,
And as an old game warden I have seen the evidence
Wrapped in layers of unholiness.

Grandmother and Grand Forests—A Southern Tale

My grandmother lived ninety-seven years. At the beginning of the
last century in their Mississippi general store, she helped her father

minister to Choctaw whose parents had escaped the Trail of Tears. She survived the bite of a "highland moccasin" as a girl and told me that her brother did so on another occasion because an aunt quickly dispatched a black hen with a single rifle shot, cut her open, and inserted the afflicted foot. Grandmother honeymooned with my grandfather sailing his skiff along wild, vacant beaches of Panhandle Florida. She knew Confederate Civil War veterans on a first-name basis. She lived so long as to witness the arrival of airplanes and automobiles, air conditioners and astronauts. Yet she did not live long enough to see the ecological restoration of forests lost in her youth. No one has.

The virgin longleaf-pine forest of the Gulf Coastal Plain (including that of Louisiana) fell as she started a family. First the saw logs and then the turpentine stumps were hauled away. As these vast parklike stands began to dwindle, the work crews moved into

Cardinal Flowers

the Pearl River swamp near her home to tackle giant hardwoods. Here most of the accessible cypress had already been floated out to downriver mills. Profits went in the opposite direction, north to nonresident industrialists who were daily reminded of the source of their wealth by the white-pine stumps in their own backyards. Local people benefited too—many of whom would have otherwise been poor and hungry. They were the ones who sharpened the crosscut saws and judged their work by the wet sound of a blade passing through three hundred annual rings in a sweetgum bole. They were the ones who nailed crescent-shaped iron shoes to the back feet of petulant oxen, and crimped dynamite-blasting caps between their incisors. My grandmother gained at the expense of these forests. Her father owned three sawmills and was considered wealthy until the Great Depression swept the mills and their money away. Later, my grandfather got a job as a timber cruiser for the railroad—and thus in an indirect way I benefited also.

Except for patchwork pastures, trees still cover most of the upland areas. It's not the same, though. Longleaf pines are too slow growing for bankers and have been supplanted by corporate loblolly. Genetically manipulated for rapid growth (and lumber quality be damned), loblolly plantations are one tier above a cotton field, biologically speaking. They should never be confused with a forest. Stands are densely planted and offer little to wildlife but sterile cover. Once vibrant players in the longleaf ecosystem, orchids, indigo snakes, and gopher tortoises have succumbed to shade, disturbance, and impatience. Those unable to adapt linger in remnant stands of marginal habitat.

After cutting, bottomland swamps were not replanted to alien species as in the uplands, and that is good. Ever resilient, they began to recover no more traumatized than by the passage of a natural and evolutionarily frequent, major hurricane. Enter then the fickle markets. Loggers went back into residual or second-growth stands to remove all red oaks for hardwood flooring, back the next decade to satisfy the new demand for ash furniture veneers and baseball bats, back the next for sweetgum required in sewing-machine cabinets and automobile bodies. All semblance of natural tree species composition and age-class distribution left with the Bachman's warblers. Yet trees remain, and as in the uplands they indeed conceal our view of a mature forest.

The terms and definitions used to describe the forests of Hernando de Soto and (375 years later) those of my adolescent grandmother are not satisfactory. Some mix economic (for example, board feet) and ecological (for example, shade tolerant) parameters in an effort to appease all. Widely hailed/hated "old-growth" roots out notions of feebleness or stagnation in my own biases. I prefer the word "mature" in the purest form. It has a feel of fullness and ripeness vital to understanding the concept. Too, my word bears a sense of cyclic naturalness that warns of the excesses of anthropocentric pondering.

The countenance of the mature forests varied with soils, elevation, and hydrology. Perhaps grandmother's swamps were the most aesthetically spectacular. One must envision a forest with many tree and shrub species, sizes, and age classes. On average, they held 75 to 150 trees per acre with 2 or 3 per acre of immense size. The age of the trees exceeded three hundred years on some sites. Forest canopies in southern swamps averaged 100 to 150 feet high and were perforated by openings that allowed light to filter down to the otherwise shaded floor. This combination of physical and temporal characteristics created a multilayered structure incompatible with commercial harvests.

Maturity in a forest is a function of age but not that characteristic alone. It must be biologically intact in that all naturally occurring species, be they wood-boring beetles, river cane, or river otters, should be present in their proper niches. Interspecific dependencies require populations to be sustainable although not static. Red-cockaded woodpeckers cannot exist if red-heart fungus disappears, zebra swallowtail butterflies are tied to pawpaws, and all predators need prey. History is revealing that sustainability is possible only within a certain threshold of human intervention. The elimination of species via direct intentional mortality is not as rampant as in the days of plume hunters and government trappers, but insidious examples are nonetheless common. Witness the disappearance of upland hardwoods in the South. Nonprofitable beech trees march away with their mysterious epiphytic beechdrops and mast-loving gray squirrels in tow. A mature forest is species rich.

Grandmother's generation developed the industrial and techno-logical ability to manipulate and change the natural world like none previous. Resulting physical and chemical alterations impede forest maturity on landscape scales. The health of her southern swamps in particular is fused to water cycles ever vulnerable to naviga-tion and flood-control projects. Even modest locks and dams raise riparian water tables enough to change a willow-oak flat to one of predominantly overcup oaks. No problem for a white-tail doe; not so for a red-headed woodpecker. Likewise, hindering natural over-bank flooding always results in a loss of forest productivity, as nu-trients are hurried downstream. A new road culvert in the swamp eliminates the ephemeral breeding pools of dusky salamanders via drainage just as a subdivision near a longleaf stand restricts those of gopher frogs by precluding fire. As for chemicals, grandmother out-lived her need to worry about malaria, although she did lament the vanishing Spanish moss in trees around the systematically poisoned soybean fields. A mature forest is progress poor.

It is tempting to sit back in this chair, an old wooden one by the way, and pass collective judgment on people who created envi-ronmental calamity—so tempting in fact that I will. The verdict is undeniable. Those folks, including kinfolks, are guilty of changing the biotic structure of this country on a scale unsurpassed since the last ice age. There will be no sentence, though, at least on that generation. All culpability is mitigated by the historian's warning to flee from the assessment of behavior in times past based on the standards of today. Grandmother would say that people didn't think of the welfare of the forests then. They needed jobs, and there was the great world war she would say. I'm fleeing.

We have gathered recently the first elementary knowledge required to put things back. It is more than planting trees, even the right ones. To achieve maturity, a forest must be allowed to resume natural cycles of succession, mortality, and regeneration. Diseases and windstorms must not always be considered unnatural harbingers of wasted fiber. Rebirth of the forest must occur in the sunbeams of openings created by fallen giants. Fauna will follow niches—big-eared bats to the hollow tupelos, nesting alligator snapping turtles to tree-fall gaps, pink mucket mussels and creole darters to unsilted streams.

On an ecosystem scale Grandmother never saw another mature, southern forest. Her generation, the last in the South to know such grandeur, has grandchildren with absolutely no hope of such an experience. For us, the grandchildren, there is no longer a need to milk every acre for its maximum economic output to keep hunger at bay and ensure the survival of our offspring. Our gift to those who may carry our genes for ninety-seven years should be the story of this loss and the resulting chance to encounter what we have missed.

Conservation Ethics

My boyhood in Louisiana was immersed in a culture where a conservation ethic did not exist in the general populace. The notion was that wildlife was there for the taking, not unlike blackberries or mayhaws in the swamp. As a carryover of attitudes about natural resources since Europeans arrived in North America, it was a lingering remnant of nineteenth-century arrogance defined as Manifest Destiny. The state of affairs was exacerbated by a legal system hobbled with weak statutes and one that did not take violations of natural-resource laws, including fish and wildlife regulations, seriously anyway. Even on the national scene there were few champions. Disciples of the great conservationist Aldo Leopold and a handful of like-minded associates were the only people touting the necessity of a conservation ethic. Most of us just did not think.

Fresh from college, I fell into the trenches of conservation ethics via law-enforcement work at my first job with the U.S. Fish and Wildlife Service at Felsenthal National Wildlife Refuge in south Arkansas. The old-school refuge manager who had just transferred in from the wilds of Alaska to start up this new federal refuge gave me and another rookie a tarnished badge and an old military-surplus .38 Special revolver stamped with "U.S. Navy." The worn leather holsters were right-handed, and I am not. With no formal law-enforcement training, we were encouraged to promote conservation ethics by enforcing all applicable federal fish and wildlife regulations. Since we loved our new charge, it did not take much encouragement. On the opening day of waterfowl season, we confiscated so many guns and illegally killed ducks on Mossy Lake that we nearly sank our overloaded boat. This was the beginning of my career as a national wildlife refuge officer, and my involvement in one aspect of conservation ethics that lasted until I retired more than thirty years later.

In the four decades since my work in the Ouachita River bottoms, illegal fish and wildlife exploitation has decreased, though it is still common enough and likely always will be. In Louisiana the more egregious violations such as gross over-limits of waterfowl, shooting wood-duck roosts, killing non-game species like night herons and robins, and market hunting have declined significantly. The change is a result of a number of factors, including environmental-education efforts in schools and by various conservation organizations. Game wardens are better trained and held to high professional standards. A maturing media has helped. The evolution of public opinion regarding poaching from tacit approval to scandalous behavior has carried over to the courts. There the status of wildlife crimes has been elevated, and penalties for violations are now severe enough to serve as effective deterrents in most cases. Changing the behavior of individuals, regardless of the technique employed (that is, proactive education or punitive mandate), has been the key to reducing violations and in turn promoting conservation ethics. Unfortunately, natural resources are besieged by commercial interests only slightly less than forty years ago. The challenge now is to get these same ethics in the doors of corporate boardrooms.

Forest vs. Trees

On these seventy-two acres where we live and that we call Heartwood I have identified 110 species of native woody plants. Over the years we purchased the property in four contiguous parcels from four different landowners. The forest on each tract has a different history according to the desires and whims of those holding title to the property. The first acquired and where our cypress house now sits is twelve acres of remarkable biodiversity. As a classic example of historic mixed pine/hardwood and hardwood slope forest, it is as close to being an intact old-growth forest of this type as exists in Louisiana. I have been told by a previous owner who was raised on the place that only a few pines were cut for stove wood decades ago, and when times were hard a few straight hickories were sold to timber buyers specializing in tool-handle stock. This parcel has the most botanical diversity and is the primary reason that the entire property was accepted into the Louisiana Natural Areas Registry.

Next in order of acquisition is the ten-acre Spring Place, once owned by my great-grandfather and named for an abundantly flowing spring. He built a small cotton gin and grist mill there that served the needs of the local community. Women would come to the spring to wash clothes, heating the cold clear water in great black wash pots. My father spoke of a large fruit orchard on part of the tract when he was a boy, and a great-uncle once had a house there. These early twentieth-century artifacts are now gone, a pipeline scars the mill site, and the spring has dried up because a surfeit of gas wells compromised the shallow aquifer. A new forest has returned and includes a scattering of old oaks and hickories that provided shade to the gin hands and washerwomen in their day.

A few years later, the twenty-acre parcel that joined the Home Place and the Spring Place became available. Its history is a classic example of land abuse. Every bole of merchantable timber was cut

Forest vs. Trees

on a recurring basis, the most recent occasion about twenty years prior. Knee-deep ruts of logging skidders still disfigured the red-clay ridges. Plant diversity existed but was suppressed by aggressive loblolly pines. Hoping for recovery in our lifetime, we bought it.

Finally, I approached the nonresident owners of thirty acres situated just north of the other three tracts and they agreed to an offer, as did my weary banker. The North Thirty added physiographic as well as botanical diversity to our now complete homestead. As a boy, my dad chopped cotton here in a ten-acre field of new ground that is discernable today only by rusting strands of hog-wire fence. Rocky Branch, a large, intermittent creek, flows east to west through a hardwood bottom and into Bayou D'Arbonne a half-mile downstream. This part of the parcel is bona fide swamp and subject to deep overflow during natural floods. Here we acquired a mayhaw patch for our jelly and a launching place for our kayaks during high water. Even so, the acreage was not without issues. Much of it had also been clear-cut twenty years earlier and planted to pines. And a crime worthy of retribution only one step removed from capital punishment involved the cypress trees. They had been girdled and likely injected with herbicide. Many survived and have recovered. Many did not.

Heartwood today consists of a seventy-two-acre block of forest in various modes of maintenance and recovery. A broad objective of our ownership is to have all of the area resemble the twelve-acre parcel, knowing full well that because we are working in tree time our grandson and his children will judge the results. Our tool kit contains drip torches for prescribed fire, chain saws for selective thinning, and death warrants for invasive species. Concerning the twelve-acre tract, we wear no blinders nor harbor misconceptions that the "good woods" will remain static. It is not the nature of any forest to remain fixed. Already drought, lightning, and dogwood anthracnose topple the Home Place favorites in the relentless trek of succession, natural and otherwise. We do, however, hope to encourage those botanical characteristics common in historic forests of this area. These include trees that are over one hundred years old, an abundance of large woody vines, distinct layering of plants (understory, midstory, and overstory), and canopy gaps that allow regeneration of shade-intolerant species such as oaks. Natural forests were also comprised of many different woody species.

One purpose of this essay is to compare the Heartwood forest to woodlands just down the road, indeed to thousands of acres of timberlands in this parish, and millions of acres in the Southeast. Up front it must be stated that these are not forests in that true forests are natural, self-replicating ecosystems that have evolved to provide nurture to an abundance of plant and animal species, including humans. Instead, these areas should be labeled inorganic, commercial tree farms. The sole objective of their corporate landowners is to produce the greatest amount of wood fiber in the shortest amount of time on the least amount of acreage. They have been tremendously successful in meeting these goals, so far.

Whereas natural forests here once consisted of more than a hundred species of native trees and shrubs, tree farms are managed for one species only—genetically modified loblolly pine. All others are discouraged by drastic means. For corporate tree farmers, starting with a scorched-earth seedling bed is most desirable. The landscape is often "burned down" with herbicides to quash competition before nursery-grown seedlings are planted in straight rows on a six-foot-by-six-foot spacing. A standard practice is to thin the plantation at twelve to fifteen years of age and clear-cut the remaining trees at twenty-five to start the process again.

Much of the land where tree farms grow is comprised of soils with low natural fertility. On these same soils second- and third-generation farmers figured this out before World War II and moved to town when their small patches would no longer support nutrient-sapping cotton. If they don't already suspect it, the tree farmers will soon discover that their intense, short crop rotation is not sustainable. Consider that the entire biomass of several hundred trees on each acre is derived only from the sun and nutrients in the soil. Consider also that new harvesting methods like whole-tree chipping returns nothing to the site, not even limbs or treetops as in traditional logging practices. Already they are forced to bomb the plantations with profit-reducing, nitrogen-based, synthetic fertilizers.

The botany of a forest is more than woody plants. In a temperate forest (and one fast becoming subtropical by definition) such as that at Heartwood, hundreds of species of herbaceous plants, wildflowers, grasses, sedges, mosses, fungi, and others fix their roots in compatible niches that aggregate into a lush botanical fabric. Animal diversity follows plant diversity, so the faunal wealth of Heartwood is affluent. Tree farms, in contrast, represent a biological desert. Three-toed box turtles that forage on our bolete mushrooms are scarce down the road, and the wood thrush whose spring song inspirits our lives could never survive in the pine monoculture. Wood ducks nest in the hollow cavities of old trees on Heartwood; any such structure that stands in the way of where a pine could grow on the tree farm is unwelcome. As I write, seven deer scoff white-oak acorns beyond my window. Deer in the pineries exist on Japanese honeysuckle and piles of corn dumped by hopeful hunters. And so it goes for the likes of buttermilk snakes, chinquapins, monarch butterflies, wild turkeys, twayblade orchids, gray fox, and so forth. A natural forest is species rich and in turn enriches the human species. Tree farms as they exist down the road are species poor, impoverishing those of us who can't see the forest for the trees.

A Political Environment

It can't be spoken in soft words for there is no other way to put it. Whether you are for it or against it, the recent sea change in American politics has led to an all-out assault on this country's long-held environmental policies and laws. In less than a month there has been serious talk of abolishing the Environmental Protection Agency. A Senate hearing was held in a first step to render impotent the Endangered Species Act. A western governor is pushing to greatly reduce the size of a recently established national monument that otherwise has tremendous public support. Wholesale elimination of government regulations including those protecting the environment is a hidebound mantra of many in public office. Barring effective pushback, a future of drastic change seems certain for our biotic natural resources.

The stated purpose of the hearing on the Endangered Species

Act was to "modernize" it. The intent, however, was revealed by the testimony of the lawmakers in control. It was declared an encroachment of states' rights, a roadblock for mining companies and other corporate interests to extract resources, and a barrier to job creation. It was said to impede land-management plans, livestock grazing, housing and commercial development. Except for the states' rights issue, the screeds possessed an element of fact. The act does indeed affect these activities on occasion. What was not considered were the numerous success stories over the life of the forty-three-year-old law. What was not allowed to be voiced in the hearing were matters of celebratory facts—that some life forms present on this planet for millions of years but approaching the brink of extinction because of human behavior still exist, indeed that some are now thriving because of this law. There was no mention of bald eagles, our national emblem that was almost poisoned into oblivion by pesticides, or California condors that completely vanished in the wild for years, or gray wolves that function as critical apex predators in complex ecosystems, species that are now flourishing under the Endangered Species Act. In Louisiana the law has facilitated the recovery of brown pelicans (our state bird), iconic alligators, and recently a unique subspecies of the black bear.

Over the course of my professional career I was fortunate to be involved in the start-up of several new national wildlife refuges. Part of my responsibilities involved writing regulations to safeguard the natural resources and the visiting public. Over time the regulations were fine-tuned, some were eliminated, and others were added to meet changing situations. But there is no scenario I can imagine that would justify the wholesale elimination of environmental regulations on principle alone. If these thoughts are upscaled to the regulatory aspects of the massive Environmental Protection Agency, I can only think that there is always room for improvement in a dynamic world and that oversight is critical. To trust though that humans, especially those that answer to profit-driven shareholders, will be good environmental stewards because of an innate sense of morality or ethical behavior is to ignore history. With only a few notable exceptions in the corporate world, regulations to protect natural resources are perceived as obstacles to short-term profits. I don't know of one significant example of a large company cleaning up its air or water pollution without threat of costly, regulatory action. The political climate today is a babelism of ideas and postures regarding the environment. Many are waving a pseudo-patriotic flag of wanton disregard of the natural world that sustains us and our progeny. Of the parties involved, only nature driving the pulsing engines of evolution is apolitical in the long run. The last word is always hers.

Perspectives

She was well over one hundred years old yet still vibrant in spirit.
Only last autumn I saw her clutching helium-filled balloons and slow
    dancing alone.

Not one to travel, she loved to hear tales of her many colorful visitors,
Especially those avant-garde transients who wintered in the tropics be-
    cause they could.

For most of her life she maintained ribald affairs with several others in the
    neighborhood.
Her paphian trysts resulted in a number of offspring, but few could survive
    in that environment.

Bawdy reputation aside (and we're talking consenting adults here),
She was always first to grant succor to generations of the needy in the form
    of room and board.

Deathly afraid of thunder and lightning as if burdened with withering
    premonitions,
She died in a spring storm after a lengthy decline.

Venerable old girl that she was, alembic in essence,
Come winter I'll be forwarding her piece by piece up the chimney to a
    Druid-filled Heaven.

∽

That a six-month-old, half-ounce bird
Gets up one morning and decides
As a matter of routine survival
To fly five thousand miles alone
To a place he's never been
On a different continent
That he mystifyingly imagines

Should humble the most egotistical of us.
That millions of such birds
Accomplish this biannual feat
As customarily as a walk
Down the driveway to the mailbox
Should render folly
And without right of appeal
Our notion of *Homo sapiens*
As some sort of supreme, tellurian species.

Deepwater Horizon Oil Spill—Revisited

In April 2010 I was swept into the wild currents of an event that has proven to be the largest environmental calamity of its type in the history of humanity—the Deepwater Horizon oil spill in the Gulf of Mexico. Not long retired, I was recruited to help assess the impacts of the ongoing disaster on Delta and Breton national wildlife refuges near the mouth of the Mississippi River. Working out of a government facility at Venice, Louisiana, we lived in an atmosphere electric with frenzied activity, excitement, and danger. At this end of the road port, hundreds of watercraft and thousands of people were being mobilized to attack an unseen enemy that was said to be approaching on distant waves. Even the president of the United States made an appearance with his imposing armada of helicopters.

With the passage of time, facts have emerged from the initial cauldron of chaos. The oil-rig explosion killed eleven people and injured seventeen others. More than 200 million gallons of crude oil leaked into the Gulf of Mexico over eighty-seven days, causing serious economic and ecologic damages to Louisiana, Mississippi, Alabama, and Florida. Thousands of birds, mammals, and turtles (some already endangered) were killed within six months of the spill. Some of the cleanup operations actually made things worse. A widely used oil dispersant was toxic to shrimp, oysters, coral, and phytoplankton. It was determined to be cancer-causing and produced mutations in fish, crabs, and shrimp. Clean-up workers were also affected by it. A subsequent study has determined that the dispersant increased the toxicity of the oil by fifty-two times. BP is now responsible for criminal and civil settlements of more than forty-two billion dollars.

Human attitudes over time regarding the disaster are especially telling. The Pew Research Center determined that support for offshore drilling plummeted after the 2010 spill. I realized the significance of this concern when visiting Secretary of Interior Ken Salazar pulled me aside in Venice to ask, "What are people saying about this?"—not "How can I help?" but "What are people saying about this?" The same Pew research noted that public support for offshore drilling had rebounded by 2015, but it also confirmed that many Americans also support investments in alternative energy. By a two-to-one ratio, a greater share of the public now favors developing alternatives like wind, solar, and hydrogen over expanding endeavors into oil, natural gas, and coal. Regardless of the road we choose, the Deepwater Horizon spill will be remembered as a calamitous obstacle in our energy path, a path that for better and worse meanders like bayous through all of our lives as well as the lives of our remaining biodiversity.

Senses of Thanksgiving

Thank you, O Lord, in this bountiful season for the five senses to relish your world.

Thank you for the succulent smells of the fruits of the earth in the kitchens of our mothers and wives. Thank you for the odor of rich delta dirt on a warm, foggy winter morning. Thank you for the smell of wood smoke, especially that tinted with lightered pine. Thank you for the stew of odors distinct to our rivers and bayous—cypress needles, primal water, mud and decay, life and life to be.

Thank you for the sound of voices of those who came before us and those who will carry our legacies into the future—our parents, grandparents, and our children. Thanks for the muffled wings of waterfowl above an overflow swamp and the belligerent snort of a doe at dusk. Thank you for haunting sounds of great horned owls and distant thunder.

Thank you for the taste of spring mayhaws and autumn muscadines in the jellies of a late November Thursday. Thank you for the abundance of other native flavors, subtle and brash—breast of teal, pecans, filet of bass. Thank you for the taste of contentment.

Thank you for the feel of a driving north wind as an Arctic front races for the Gulf. Thanks for the textures of sweet gum balls, feathers, gumbo clay, and beech bark. Thank you for the heat of an open fire and the warmth of an open heart.

Thank you for the sight of falling leaves, fattening squirrels, and rising waters that foretell the change of seasons. As the sun approaches the solstice, thank you for lengthy shadows and longer sunsets. Thanks also for fleeting glimpses—of a bobcat at dawn, of a shooting star on a rawboned night, of curiosity on the face of a young grandson.

I pray also, O Lord, for a sixth sense. Grant us common sense to be good stewards of these treasures. Amen.

INDEX OF FLORA AND FAUNA